Discovering the Thousand Islands

The Thousand Islands

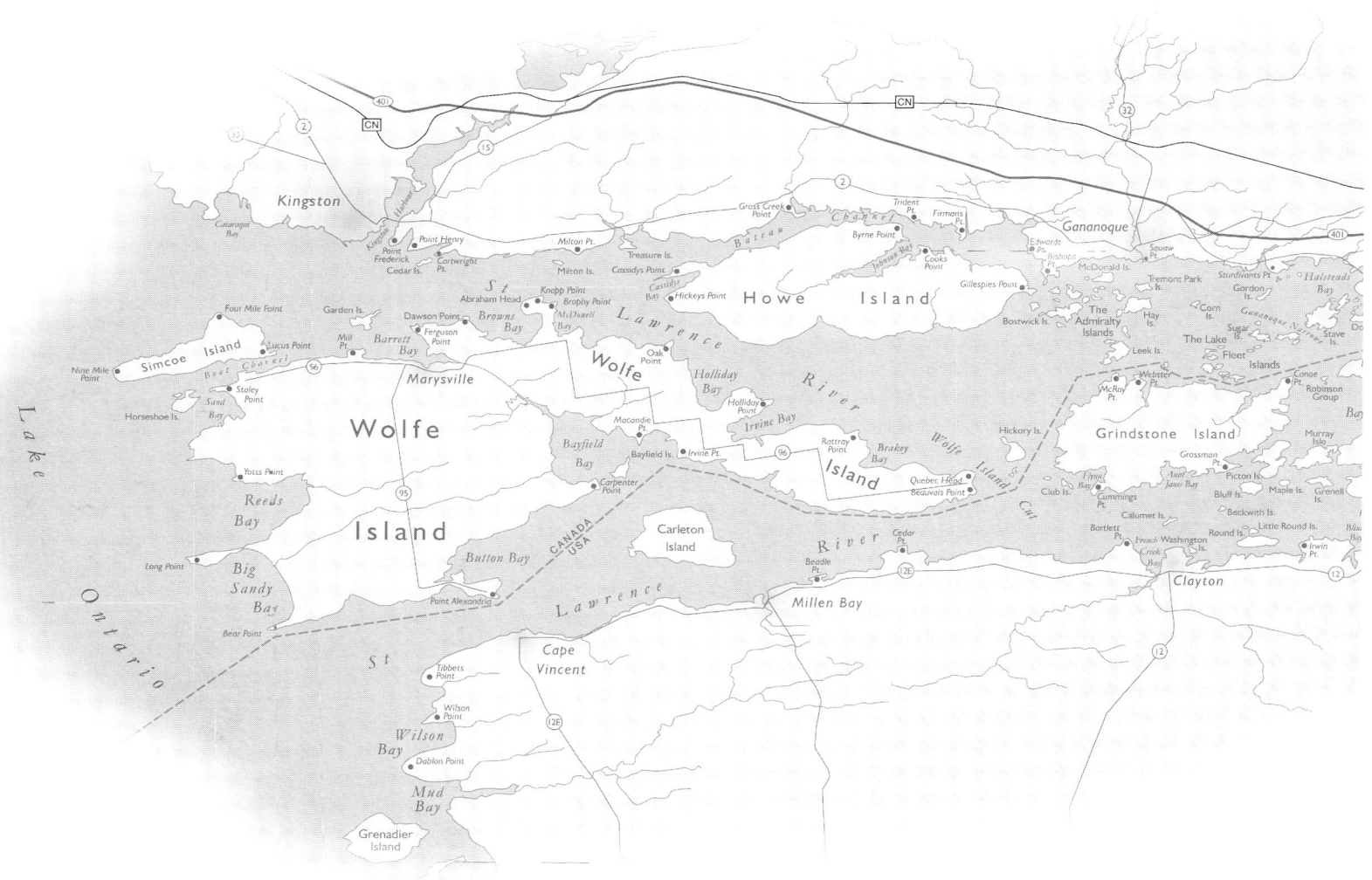

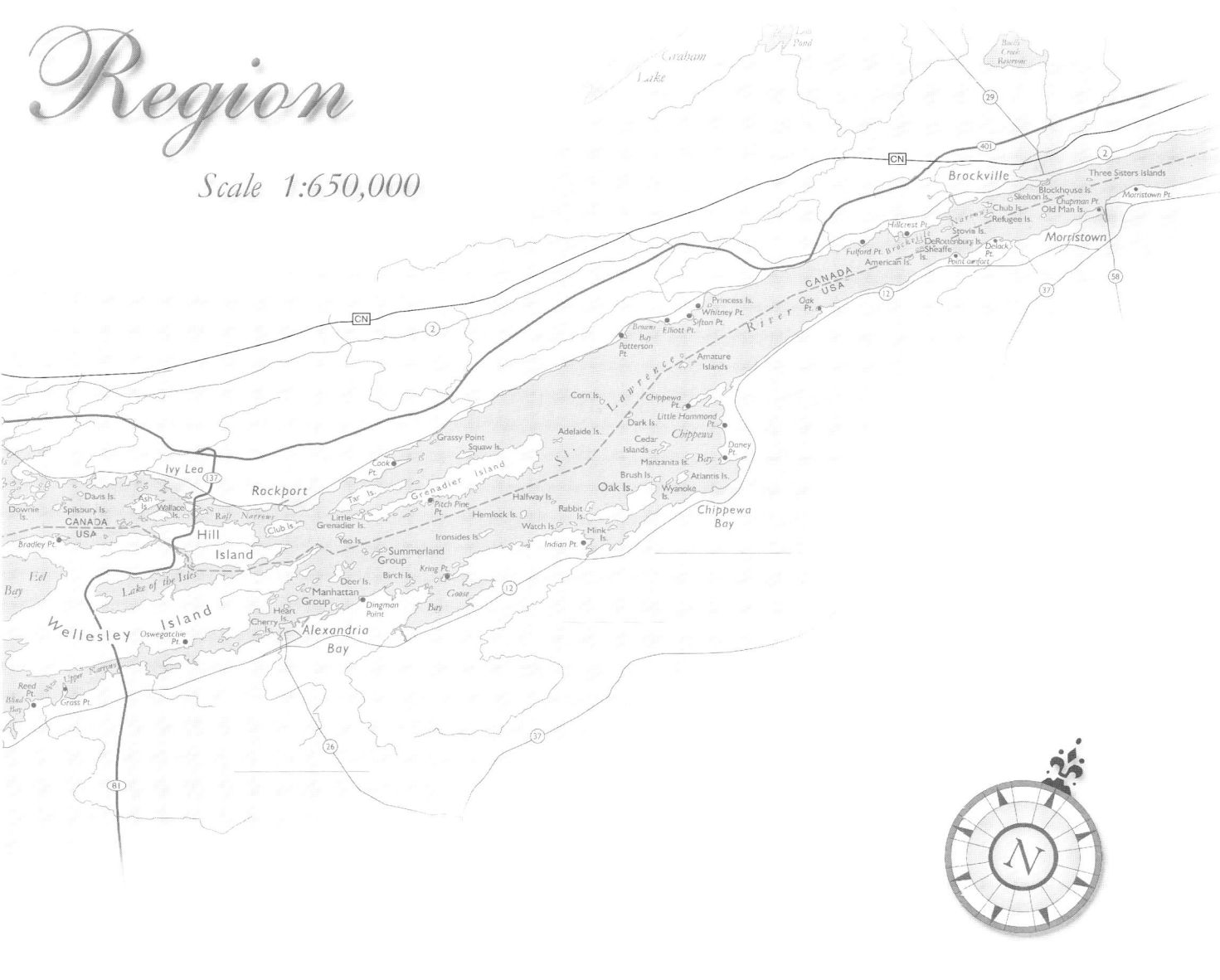

Region

Scale 1:650,000

Summer pleasures of the Thousand Islands — cottages, boats, and refreshing swims in the St. Lawrence River. (Antique Boat Museum)

*D*iscovering the Thousand Islands

Don Ross

QUARRY
HERITAGE
BOOKS

Copyright © Don Ross, 2001.
All rights reserved.

The publisher acknowledges the support of
the Government of Canada, Book Publishing
Industry Development Program,
Department of Canadian Heritage.

ISBN 1-55082-285-3

Design by Susan Hannah.

Printed and bound in Canada by Custom
Printers, Renfrew, Ontario.

Published by Quarry Press Inc., P.O. Box
1061, Kingston, Ontario K7L 4Y5 Canada,
www.quarrypress.com

ACKNOWLEDGEMENTS

Many very special people contributed to this book, all helping to steer the pen in one way or another. Sometimes, ideas and inspiration are like errant waves that splash in over the transom of your boat: unexpected and often eye-opening surprises. People on the river occasionally drop wonderful tales and priceless nuggets of information in the middle of conversation. There is as well the circle of friends and family that help define those important and pivotal moments of life in the islands. I'll always cherish the memories of wonderful times on boats and islands with my dearest wife Marni and her lifelong friends, the Merrimaids — a foursome whose singing formed the bond that often brought them and their husbands together for island holidays. Reflecting on those days together, I turned from resident storyteller to scribe. And last, but not least, a warm thanks to my dear daughter Jess who steered this project straight and true with a firm grip on her style guide and clever words of encouragement.

All photographs are by the author or from the author's collection, unless otherwise credited. Maps by Virginia Adams. For research and illustrations, special thanks to the Antique Boat Museum and Phoebe Triton, Clayton, New York; Verda Corbin, Les Corbin Studio, Clayton, New York; Parks Canada, St. Lawrence Islands National Park; National Archives of Canada, Ottawa, Ontario.

CONTENTS

FOREWORD

Garden of the Great Spirit

The sun breaks through thunder-clouds in the Thousand Islands.

I n a rage of disappointment, the Great Spirit sent his warriors forth carrying an enormous animal hide to scoop up the Garden of the Great Spirit. The Great Spirit Manitou had created a heaven-on-earth as a reward when the native nations at last agreed to stop fighting among themselves. The garden was a wonderful land of plenty, where game abounded and summer never ended, but this peaceful paradise was enjoyed for only a short time before war broke out again. If these ungrateful nations could not live in harmony, the Great Spirit declared, then they would go back to the old hardships of the rugged land! But just as the garden was being ripped from the earth, a corner of the hide broke free. Some of the rocks spilled back to earth, and a thousand fragments of paradise remained where they fell, scattered in the river as the Thousand Islands.

Among the first peoples of the region, the Algonquin and Iroquois nations, there are many creation myths like this story of the Garden of the Great Spirit to account for the breathtaking beauty of the Thousand Islands. Another legend tells of two powerful gods, one good and the other evil, who argued fiercely over who would rule the mighty St. Lawrence River. The argument escalated into an earth-shaking combat when each god tore huge handfuls of rock from the face of the earth to heave furiously across the river at the other. Fistfuls of rock fell short of their targets on the shores, landing in the river. Finally, the good god triumphed, and evil spirits were forever banished from the land. The good god cast a spell on the hundreds of chunks of rugged rock scattered in the river, and enchanted deep green forests flourished there. These thousand verdant islands became known as Manitouana, the Garden of the Great Spirit.

The figure of Hiawatha plays a role in another Algonquin or Ojibway myth. The story tells of a battle fought between a father and his son, the West Wind and Hiawatha, where Hiawatha sought to avenge the death of his mother at his father's hand. The West Wind confessed to Hiawatha that his greatest fear was of the 'black' rock, those dark veins in the granite found throughout the islands. Seeing an advantage, Hiawatha lied, saying that bulrushes were what frightened him most. When the fight started, Hiawatha hurled black rock boulders across the river at his father, who tried to defend himself by heaving back ineffective mounds of bulrushes. When the fight had ended, the islands and marshes of the Thousand Islands remained as a legacy of the battle.

Among the Iroquois, there is a legend of Ta-oun-wat-ha who was once sent by the Great Spirit to find the most beautiful place on earth to give to the Five Nations of the Iroquois. He came to the Thousand Islands in a white canoe, and after exploring the islands, he knew that this was indeed the place, but it would take just a bit more work to make the region perfect: he spent the next few years making deeper channels and splitting rock ridges into the cliff faces we see today. When Ta-oun-wat-ha gave up his otherworldly position as a deity and settled on earth to be married, he became known as Hayo-went'ha, sounding in name like the Ojibway god Hiawatha, but not the same entity.

The first peoples who discovered the Thousand Islands are not the only ones to have been inspired to praise their serene yet spectacular beauty. Centuries later another great mythmaker, the American poet Walt Whitman, would explore those deeper channels of the Thousand Islands by steamboat and proclaim the spiritual grandeur of "the most beautiful extensive region of lakes and islands one can probably see on earth . . . The beauty of the spot all through the day, the sunlit waters, the fanning breeze, the rocky and cedar-bronzed islets, the larger islands with fields and farms, the white-winged yachts and shooting row-boats, and over all the blue arching sky copious

— make a sane, calm, eternal picture, to eyes, senses, my soul. . . . Land of pure air! Land of unnumbered lakes! Land of the islets and the woods!"

The Thousand Islands is truly an awe-inspiring, soul-satisfying place, a place where the rock is etched by lines of time as deep and as old as a quarter of the age of the earth and as long as the history of any living thing . . .

. . . where pale rose and tan granites, the roots of ancient mountains lifted from the very core of the earth, lie in rugged chains across the young river, and sieve the flow of the St. Lawrence on its passage to the sea

. . . where through the aeons, the rock has shrugged off glaciers, the scouring river, the best and worst of weather, the timeless passage of the seasons, and the earthy litter of mosses and bushes and trees and tangles of thickets that cling to its hard, enduring surface

. . . where the cliffs have seen all of history and have heard every sound of man, echoing every voice of discovery and settlement, every struggle and conflict, and all sounds of holiday pleasure

. . . where minnows flash silver over skiff-shadowed shoals and small boys swing on creaking ropes slung from overhanging trees and plunge laughing to the emerald cool of the river

. . . where the golden-red sun of day's end slips behind the perfect black silhouettes of wind-flagged pines and turns the mercury-smooth river ripples to mauve and magenta and the color of hammered copper

. . . where night is a blue-black velvet veil, pulled across the sky by the vanishing sun and silver pierced by all the stars of the universe

. . . where the mists of summer mornings float like layered cotton gauze between the soft wet green of neighboring islands

. . . where the mists of frozen winter mornings twist and spiral and dance above the river's lingering warmth, gathering overhead in small puffs of cloud or settling like crystal feathers on winter-brown shoreline grasses

Looking out to Fiddler's Elbow Channel,
where fiddle player Chauncy Patterson
entertained passing tourboats in the late 1800s.
(Antique Boat Museum)

. . . where great blue herons claim choice rocks and overhangs in sun-warmed bays, 'quwarking' and 'quworking' and swearing 'heronese' protests and insults at intruders before winging their graceful gangling way to the solitudes of a neighboring haunt

. . . where every hue of green is painted in the woodlands and plays in the reflections on the water and a thousand grays flavor the mists of morning and fogs that follow on the heels of summer storms

. . . where impossibly brilliant greens of mosses, wildflowers and unfurling foliage of trees are the vibrant backdrop for every color of spring flower

. . . where even in winter, the frozen blue-greens of the woodlands and wildland shores enliven the somber rock and the dormant brown forests in their cold blue-and-white slumber

. . . where in the cold forests of winter, lofty, frost-hardened branches of ash, aspen, and maple sway and slap and rattle, almost drowning the droning whispers of wind that sieve through the needles of the pines

. . . where in the fabulous warmth of the summers, those same limbs swirl lazy and almost soundless breathes of soft green haze against the cathedral of the sky and the pines murmur monotone sighs of content

. . . where a thousand shapes of waves ripple and tumble and roll their steel gray and emerald green and sunset-copper crests toward the frost-split, broken rock of the shores

. . . where the island spices of resinous pines, sun-baked mosses and blueberries, and dusky hemlocks bend over the river-cooled water to brush your face as your boat drifts on sun-sparkled wavetops through ever-winding and endless channels. . .

This is the Thousand Islands, the Garden of the Great Spirit, ready to be discovered by each succeeding generation of islanders and visitors.

1

Bearings & Boundaries

A THOUSAND ISLAN ... AND COUNTING

The Thousand Islands International Bridge.

W here on earth is the Thousand Islands? On your atlas or globe, look for the Thousand Islands centered roughly at 76 degrees longitude and 44.4 degrees latitude, about a fifth of the way inland from the east coast of North America, and about halfway from north to south on the continent. The region lies south of most of Canada, south even of most or all of the American states of Washington, Montana, North Dakota, Minnesota, and Maine. When viewed from Europe and Asia, the Thousand Islands is also south of all of Great Britain, Germany, Switzerland, Austria, and the largest part of France, at the same latitude as the Bordeaux region of France, the Italian city of Genoa, Bucharest in Romania, and the highlands of the island of Hokkaido in Japan.

The Thousand Islands region begins in the west where Lake Ontario flows into the St. Lawrence River, with Kingston, Ontario to the north in Canada and Cape Vincent, New York to the south in the United States. To the east, the region is bounded by Brockville, Ontario and Morristown, New York. Two major highways lead to the region. Interstate 81 runs

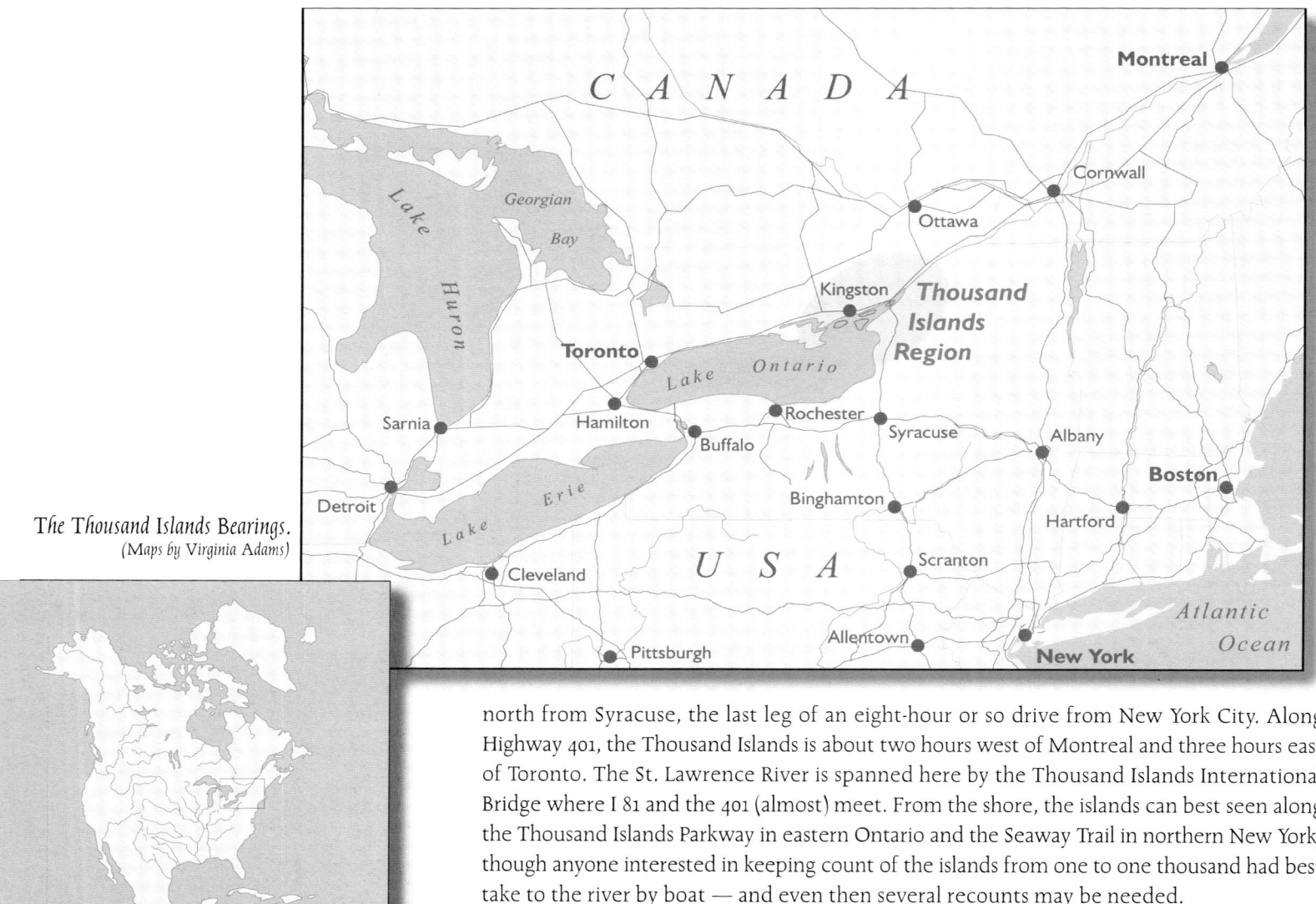

The Thousand Islands Bearings.
(Maps by Virginia Adams)

north from Syracuse, the last leg of an eight-hour or so drive from New York City. Along Highway 401, the Thousand Islands is about two hours west of Montreal and three hours east of Toronto. The St. Lawrence River is spanned here by the Thousand Islands International Bridge where I 81 and the 401 (almost) meet. From the shore, the islands can best seen along the Thousand Islands Parkway in eastern Ontario and the Seaway Trail in northern New York, though anyone interested in keeping count of the islands from one to one thousand had best take to the river by boat — and even then several recounts may be needed.

Just how many islands are there in the 'Thousand' Islands? Believe it or not, after all this time, there is no definitive answer. The number depends on how one defines an island, when in history the count took place, and even in what season the count took place.

Webster's Dictionary states an island is any piece of land, smaller than a continent, that is entirely surrounded by water. Another definition, often bandied about on the tourboat circuit, is anything with six square yards of grass and two or more trees. Is an island an island if it's big enough to stand on, or must it have a certain amount of grass, regardless of the number of trees, bushes, thickets of wildflowers, or mounds of moss? With criteria that loose, there is certainly room for variation in counting the number of islands in the river.

To compound the conundrum of the count, the level of the water in the river varies well over a meter through the course of a year. It is highest in early summer and lowest in mid winter. Rocks that are awash in the high waters of June and July may have become large, dry islets when the water drops in the late fall. When surveys were carried out in the Thousand Islands in the early 1800s, dams had not yet been built downriver to regulate the river's flow. Water levels fluctuated considerably more through the seasons than they do now. In fact, some islands surveyed in the fall and winter were recorded at up to a fifth of an acre, but by summer, even perhaps by the time the early surveyors had moved to another section of the river, they were completely submerged. In the old days, some of these 'wet' islands were actually purchased on good faith by speculators. Taxes were paid for decades by their unwitting owners and even their heirs — if they didn't take the trouble to visit their 'island' property. Some of these official islands have only been dry in years when the water was at most unusually low levels.

Other islands have been joined to the mainland with loads of fill as causeways, creating points of land. A few islands were blown out of existence in the building of the St. Lawrence Seaway. Some of the particularly tiny islands have lost their claim to island status when their the precarious two small trees were lost to especially violent storms. Yesterday's island becomes today's shoal. There are many other places where dense cattail marshes have filled in around numerous small islands in shallow water, disguising their true nature. Some islands near shore have been linked by fill for roadways to the mainland so that they now appear as points of land.

So how many islands are there now? When early French cartographers named of the region Les Milles Isles, or Thousand Islands, they underestimated. The number is about 1,830, though the most careful count on the charts will not reveal so many. The charts are excellent for navigation, but at their scale, quite a number of very small islets appear joined together.

How to best count the islands? From a small boat, with charts and note pad and plenty of unhurried summer afternoons.

White-painted, thigh-high concrete boundary markers were placed on some island shores where they lie close to the international border.

SURVEYING AND NAMING THE ISLANDS

I f the count of the islands has varied over the years, so have their names in many cases. Almost all of the islands have names — and many have had two or more over the years. No doubt, there were many more names given by passing travelers through the centuries that have been forgotten, misplaced, or displaced as newcomers labeled the islands for themselves. Islands were named for people who owned them, events that happened on them, and special features about them. There was a time, though, when a major effort was made to survey and name all of the Thousand Islands.

Immediately after the end of the War of 1812–1814, it was clear to both the British and the Americans that there was a real need for a proper survey and mapping of this region to draw a well-marked boundary between Canada and the United States. At the end of the war, the British Admiralty chose Sir Edward Campbell Rich Owen, commander-in-chief of the navy on the Great Lakes, to wind down the fleet, the shipyards, and the size of the navy to the point where the force could still be effective if need be, but to a cost that the Admiralty could bear. In 1815, the Admiralty also sent Edward's brother, Captain William Fitzwilliam Owen, to Kingston, charged with the task of surveying the St. Lawrence River through the Thousand Islands and the first three of the Great Lakes.

The monumental task of mapping the Thousand Islands began in the winter of 1816. For the first time, careful measurements were taken to reference the exact position of the islands and shoreline. Depths of the channels were sounded as well. The survey crews broke new ground in their methods. Distant landmarks were triangulated by sextant and compass readings that pinpointed places from which signal rockets were fired. Notebooks were taken back to the hydrographer's house in Kingston to be carefully transcribed to preliminary maps. While the shapes drawn for the islands were not always precise, their relative locations were. In addition, these first charts contained a wealth of other information, including directions of currents, forest cover, marshland, and much more. Although there would be new surveys to aid in fixing the boundary line and to clarify the shapes of the islands, Owen's work was a wonderful benchmark.

The charts used in the Islands today have evolved from Owen's survey, and many of the island names harken back to the original survey. A good part of the reason for this is owed to Owen's clever insight into the landscape of the Thousand Islands. The islands do not lie at random in the river. Rather, because they are the roots of ancient mountains, they form a chain-like pattern that happens to run at a slight diagonal across the flow of the river. After all of the islands were plotted, it struck Owen that the clusters and chains of islands lay in the river like fleets of ships. This presented the opportunity to name the island clusters after persons and ships of the period. He didn't confine himself to just the British territories, but named islands and fleets of islands throughout the region. Modern charts still name most of the original thirteen theme groups seen by Owen.

The Admiralty Group off Gananoque celebrated persons within the British Admiralty, including lords and senior naval officers, some of whom were responsible for provisioning and supply and so would have been especially important to the surveyors. Owen liberally sprinkled names like Hurd, for an Admiralty hydrographer; Walker, for a chart draftsman; Yorke, for Sir Joseph Sydney Yorke who was an Admiralty commissioner and rear admiral; and Warrender, for a fellow commissioner of Yorke. However, anyone who looks at today's charts of the region will find very few of Owen's dedications in the Admiralty Islands. Instead, most of the islands have been renamed over the years for local people and characteristics of the islands themselves. For example, Hurd became Buck, after a lightkeeper, and then Beau Rivage about the time it was included in the St. Lawrence Islands National Park, which itself dates from 1905. Walker Island became Mermaid, again as a park island; Yorke was changed to Bostwick, apparently adapted from Boss Dick, a work foreman on the island's granite quarry; and Warrender gave way to McDonald, a hog farmer who used the island in summer. It's not hard to see where some of the other islands, such as Pitch Pine, Burnt, Hemlock, and Hay, got their local names. The Admiralty may have been honored as a group, but the individuals may not have been impressed if they had visited years later.

The Canadian Span of the Thousand Islands International Bridge.

The colorful island names of Deathdealer, Dumfounder, Bloodletter, Scorpion, and Psyche are after British ships on the lakes during the War of 1812. These islands are in the Lake Fleet, south of the Admiralty group. Many of these names have remained untouched, probably because they are so bold and interesting. The Navy Fleet — Mulcaster, Popham, Spilsbury, and Cunliffe — recognizes ships' commanders in the war. Owen placed his own name on an island amongst these. There are local names scattered here as well, but among the other names that stuck are Turnip, Potato, and Cod Fish — no doubt, something of a stew.

The Amateur Group, between Mallorytown Landing and Brockville, remembers a group that was outside of the navy itself, and hence 'amateurs', engineers and soldiers likely to have been known by the cartographers. The Brock Group, like Brockville itself, was named after Sir Isaac Brock, a heroic commander who fell in the War of 1812, and for other British navy and army officers.

In her book *The First Summer People*, Susan Weston Smith traces the long, detailed, and fascinating origin and evolution of these names.

BORDER LINES

When it came to actually placing the boundary between Canada and the United States, it was pretty obvious that it wouldn't be as simple as running a line up the middle of the river. The river is full of islands, and a straight line would lie across and divide a number of them. This would create headaches in the future. If a boundary is to help define jurisdiction and rights for commerce, such a line across islands and straight up the river would only lead to fracturing the law. That realization prompted an article of the Treaty of Ghent, signed to end the War of 1812–1814, to undertake a workable definition of the "middle" of the river. Two groups of surveyors were assigned, one American and the other British. They provided the information that commissioners would need to negotiate the boundary and give, as much as possible, equal ownership of the islands to each country. The work surveying the border and allotting the islands took a great deal of time lengthy negotiations. The final agreement was penned in 1822.

The result is a boundary that tracks in and around the islands, but touches upon no island in the process. The border line doesn't necessarily follow main channels. In fact, there are places such as the International Rift between Hill and Wellesley Islands where the passageways are so narrow that it seems one can almost jump from one country to the other. The agreements gave two of the four largest islands to Canada — Wolfe and Howe — and the United States acquired Grindstone and Wellesley. The remainder of the islands were apportioned as equally as was possible, given the challenge of putting the boundary up the 'middle' of the river.

Postcard of Captain Snider's tour boat in a Canadian channel.

BORDER STORIES

The fact that the border between Canada and the United States weaves among the Thousand Islands has played well for tourboat commentators over the years.

With tongue in cheek, one lectured that the boundary marking was considered a serious matter between special commissions set up just for that purpose. "Each winter," he would intone, "supervised works crews would go out on the frozen river and re-survey the exact location of the line. They would then paint that line with a fresh coat of red lead paint. When the ice melts in the spring, that line sinks to the bottom of the river." And here he would pause for effect and walk to the side of the boat and look down into the water. "As a matter of fact, we're about to cross that line now." Regardless of how improbable the story seemed, everyone would have to see for themselves … and then have a good laugh.

Another story tells of a captain of one of the small, wood-planked tourboats that used to carry passengers until the 1960s, when much larger vessels began to take over the routes. This gentleman played upon the notion that Canada is the colder of the two countries. There was a point in his tale where he would announce that the boat was now traveling right along the international boundary, with Canada on one side and the United States on the other. The captain then invited his passengers to dip a hand in the river, first over one side of the boat and then the other. Now, as it happened, the water-cooled engine was fairly far forward and exhausted on one side of the boat, just above the water line. The engine-heated water flowed back along the hull and so the water was noticeably warmer to the passengers on that side, which happened to be the side facing the American shore. It could only be true, then, that Canada was the cooler of the two countries!

Yet another borderline story goes that the international boundary passes between Zavikon Island, a little south of Rockport, and the small island to the south of Zavikon. There is an arched bridge linking the two islands, and flags painted on the bridge and flown from the islands suggest that the main island is in Canada and the other is in the United States. Again, tourboat commentators have seized on an opportunity and relate that when the island's owner grows tired of listening to his mother in law, he can send her across the bridge and out of the country. As anyone who takes a look at a river chart can see, the story has one big flaw: the true boundary line actually lies well south of both islands. The flags are flown just for effect.

2
Landscape & Wildlife

A STORY IN STONE

*Root-fractured rock
at Camelot Island.*

At the dead of winter in the middle of the night, the Thousand Islands can be so very still that nature seems to hold her breath. There are times when the temperature plunges so low that the night air seems as hard and dry as frozen steel. Then everything seems to startle when a snow-muffled 'crack' thunders from a granite ridge, shattering the stillness of the forest night. For summer after summer, filament-fine rootlets of a white pine, birch, or hemlock have worked their way into a hairline crack in what seems to be flawless, smooth granite. The prying roots wedge the surface crack a little wider each season, letting in first a trace of moisture and then a trickle of rain. That's all it takes. In the bitter cold of the winter's night, the frozen water becomes an irresistible force, shattering the root-wedged face of granite, cleaving it as cleanly as if by a stone cutter's hand. A face of rock divided, a boulder split by frost, a shoreline stone fractured by winter ice, pebbles tumbled together in the rhythm of waves, big and small, on the beach … Countless times every day, every season, seemingly indestructible stone weathers a little more. The stuff of mountains is finally reduced to silt, some to rest on the river floor, some off on a journey to the sea.

The story of the Thousand Islands begins here, long ago when the continents we recognize today were still forming, long before any but the most primitive living thing graced the face of this planet. Nearly one billion years ago, the continents had only just begun to form. The part of the world that was to become the Thousand Islands lay at the seabed, near the edge of the forming continent of North America. Weathered grains and particles of rock which were sloughed off in the turbulent formation of this planet settled on the ocean's bottom. They collected in layers which in time became deep enough to form rock. Then, because of the relentless forces created by the constantly drifting crust of the earth, the seabed slowly buckled. Some of the folds were pushed far down into the crust of the earth and some were tilted skyward. Enormous chains of mountains shuddered, quaked, and 'volcanoed' into being as this block of the North American continent emerged from the sea.

The sandstones thrust deep into the earth were metamorphosed, physically changed by the awesome pressures and heat. They were squeezed and melted into new and very hard rock forms called schists and gneiss. Fissures in these rocks were penetrated by molten granite and basalt from deeper in the earth, leaving seams of different colors and textures than in the surrounding rock. Earthquakes continued to shift and shatter the rock in the great depths below the mountains, often causing slabs of rock, great and small, to shear apart and become offset. This created the mismatched puzzle-piece effect that we see in exposed rock faces today, where the various colors of seams and lines in the rock are unaligned by a few centimeters or even by meters.

Every mountain is wearing down, even as it is being built. The seabed sandstones aloft at this new roof of the world faced punishing weather. Despite the sandstone's rough, hard surface, it was no match for the water and wind, freeze and thaw that wore grains, pebbles, and even slabs from the mountain sides. Gravity nudged each bit to lower levels, runoff urged each fragment seaward once again. After hundreds of millions of years, the peaks became rolling hills, and the hard roots of the old mountains finally rose to the surface of the earth. Weather and time took their toll, too, even on these durable granites. The landscape that evolved, except for quite different vegetation, looked very much like the one seen today.

But then change on a massive scale began again. The part of the world that was to become the Thousand Islands gradually sagged below sea level and lay submerged for hundreds of millions of years more. Over that time, sediments of other weathering landscapes accumulated atop the mountain roots. In the great depths, these seabeds were pressured into sandstones and limestones, this time entombing the burrows of primitive worms and the shells of sea life, hinting that this part of the continent had drifted to tropical latitudes.

These younger sedimentary rocks still form the bedrock of the landscape of much of the lower Great Lakes, but not in the Thousand Islands. Renewed pressures in the earth lifted a long ridge at the southern edge of the Canadian Shield, reaching as far south as another uplifted region, the Adirondack Mountains. Weathering once again helped slough off the overbearing softer rock from the old mountain roots. When glaciers of four ice ages bulldozed their way south and west across this part of the continent, nearly all of the younger rock was scraped from this south-reaching ridge of the Canadian Shield. The landform called the Frontenac Arch had come into being. When the ice of the last glacial age melted away some 10,000 years ago, and after the land gradually rebounded when released from the incredible weight of the glacial ice, today's Thousand Islands landscape emerged.

Virtually all of this geological history of the Thousand Islands can be read on the surface of the land. Twisted swirls of black and gray rock are the time- and heat-tortured remains of nearly billion-year-old sandstone and limestone formations from the deep base of mountains that lofted here 600 million years ago. The pink granites are fractured and shot through with milky veins of quartz and dark, dull, gray lines of basalt from the depths of the earth, pushed up under the base of the mountains. To the east and west of the Thousand Islands, and even on some of the islands themselves, there are layered sandstones and limestones that formed over the weather-worn roots of the old mountains 300 to 400 hundred million years ago. Etched on the smooth slopes of the granite are the marks of the passing of the glaciers: shallow, rounded furrows ground out as ice pushed boulders and gravel along, and crescent-shaped 'chatter marks' chipped from the surface as hard rock was dragged and bounced along by the ice.

There is also plenty of evidence that points to the processes of the continuing evolution of this landform: frost-split boulders and stones that rim the island shores; root-wedged flat faces of rock; random piles of broken rock laying where they fell from the grip of the roots of fallen, long-since rotted trees. Pebbles and grains of sand are rinsed

Basalt intrudes ancient fractures in island granite rock.

into the river by runoff and tumble in the waves along the shore. Some day, worn to silt, they will drift on the currents to the depths of the sea.

ICE AGES

The ice ages in North America were the brush that wiped the slate clean. Each of the four glacial periods had such a powerful force that they did more than freeze the life on the land, they obliterated it. At the end of the last glaciation, the Wisconsin period, southern Ontario was freed from its enormous burden of ice and an environment similar to today's northern tundra gradually began to evolve. The scouring force of the glaciers had virtually stripped the soil of organic matter, leaving this region as a desolate landscape of exposed rock, gravely ridges, and barrens of silt and clay.

The Great Lakes came into being when glaciers gouged out the basin where they now lie. Before the ice ages (perhaps right up until after the last one), the lakes were a series of lesser lakes and rivers which emptied southward into the Ohio and Mississippi River valleys. The Frontenac Arch, the backbone of the Thousand Islands, was an uplifted ridge even before glaciation that all but prevented the forerunners of the Great Lakes from flowing east to the Atlantic.

Weathered and glacier-sculpted granite on the north-east shore of Wellesley Island.

During the last ice age, the stupendous weight of ice over a kilometer thick depressed all of the land, including the Frontenac Arch and the basin of the newly formed Great Lakes. In fact, the land was pushed down so much it went below sea level and the ocean flooded as far inland as the Thousand Islands. This arm of the Atlantic, the Gilbert Gulf, allowed saltwater fish and sea mammals, such as the beluga whale, to inhabit the region briefly, and created sand beaches and dunes in eastern Ontario and upper New York State. As a result of this ocean invasion, a corridor opened from the

Atlantic to the upper St. Lawrence for colonization by coastal species, such as wire birch and red spruce that still grow here, but at their western range limits.

The meltwaters from the glaciers found their way seaward as the land gradually rebounded. The shores of the post-glacial lakes stood at different levels than are seen today, but there is clear evidence in the Thousand Islands of where one shore sat for what must have been a considerable period of time. On virtually every steep rocky shore, about four to eight or so meters above today's river level, there is a broad band of coarse, broken rock. This is where shoreline ice fractured boulders in the same way that happens each winter along present-day shores. There are also many places where the silty clays of the ancient lake bottom settled in rock-lined valleys. These are found at a level a little below the old broken rock shores. If shore erosion in recent years has cut back into such clay basins, layer upon layer can be seen in the clay, just like growth rings in trees, where the coarsest particles settled in the open waters of summer and the finest particles were able to settle out in the calm water under winter ice.

ICE ISLAND

Hunter's refuge and stone duck blinds on Ice Island where winter ice floes break up.

Spring break-up in the Thousand Islands is one of the most powerful statements of the river's force. It is not usually until early January that the massive flow of the river yields to the forming ice, and in a succession of cold, still nights, the quieter channels finally freeze in. Aside from when winter winds reach such violence that surging waves heave the frozen surface into fragments and floes, much of the river, except channels where the swift flow never yields to the ice, is ice-locked for the duration of the winter. The river's surface becomes motionless until the spring sun weakens the sheets of ice.

It takes weeks for the deep-frozen ice to clear the channels. Great chunks crack free of the mass and drift with the current.

The steady rumbling and shatter-crashing flow of breaking ice seems endless, but the Thousand Island ice comes mostly from the river; only a little bit from Lake Ontario. The ice is borne on the current, turned in eddies, crushed on the edges by collisions with other floes and shorelines, and is scarred and fractured by shoals and navigation spars. But there is one place on the river where the drifting ice is very dramatically heaved and smashed, if it is carried to that spot.

Ice Island is a small, narrow, and treeless island in the Canadian Channel, just east of Mallorytown Landing and Grenadier Island. From the Thousand Islands Parkway, it would fall into the same view as Dark Island's Singer Castle. Ice Island's western end rises from the water in a low, wedge-shaped point of granite, a little like the prow of a ship. Every spring, the west-facing point of hard granite endures a mini Ice Age of sorts as the irresistible flow of the river carries floe after floe, large and small, thick and thin, onto the point of the island. Any loose rock has long since been carried or pushed away. The island is like an ice-breaker that waits for the ice to come to it. In some years, after a particularly long and cold winter, the sheets of ice that are stranded upon the island pile to such heights that they may endure well into May, months after the rest of the winter ice is a memory.

The Thousand Islands are at an ecological crossroads, where the Great Lakes escape oceanward down the St. Lawrence River through a low part of the Frontenac Arch.
(Map by Virginia Adams)

ECOLOGICAL CROSSROADS

From about 12,500 to 10,000 BP ('Before Present', an archeologist's term), the winters were cold and the summers were cool. The tundra-like environment was slow to give way to the boreal type forest in those conditions. Finally, around 12,000 years ago, things began to warm up a bit.

Nature is always quick to fill a void. Meadows of dwarf willows, stunted spruce, chickweeds, sedges, and mountain avens clung to the barren soil. Ponds and streams were colonized with pondweeds, millefoil, and watershield. An overland invasion of species took place around the shores of the receding Champlain Sea from the New England area, the Appalachians, and the Adirondacks. Spruce became the predominant tree of the forest by around 10,000 BP, but its presence enriched and sheltered the ground, paving the way for red and jack pine forests. As the height of the trees grew, the water-holding ability of the soil increased because of the shade. Plant growth and therefore soil development accelerated. Over the next 2,500 years, the climate became a little milder and the soils a little richer. White pine became the dominant tree of the forest. A bit more time still and deciduous and hemlock trees added to the ever-enriched environment.

The stage was now set and the conditions were about perfect for the Thousand Islands to become one of the most ecologically rich regions on the continent. The long geological history created a rugged, complex, and extremely varied physical environment. There are rock outcrops, pockets of heavy clay, sand and gravel ridges, well-drained soils, seasonally flooded fens, cliff faces — and plenty of variations on all of the above. There are many types of bedrock exposed to lend to the changes in the soil chemistry. Granites, sandstone, limestone, gneiss, and schist all influence the plants that can grow atop them.

The region is at an ecological crossroads, of sorts. The north–south road is the Frontenac Arch, the uplifted granite roots of the ancient mountain chain connecting the Algonquin Canadian Shield region to the rugged granite hills of the Adirondack Mountains. The east–west road is the St. Lawrence River. The Frontenac Arch pushes up through the much younger and softer sedimentary bedrock of the lower Great Lakes and St. Lawrence valley, so prominently that it almost blocks the oceanward flow of the Lake Ontario. As the waters of the lake sieve through the hills of the Arch, the St. Lawrence River begins. From this route, the Thousand Islands take on some of the character of the Canadian Maritimes, New England coastal areas, and Appalachia; and some of the character of the prairies and plains along the Great Lakes basin that was no doubt the route taken by some of the southern flora found here.

What these corridors do is provide the routes for colonization and migration of some plants and animals, while presenting barriers to others. The Frontenac Arch links the similar geological formations of the Canadian Shield to the Adirondack Mountains. A narrow structure, somewhat like the narrow neck of an hour glass, the Arch forms a con-

Deerberry, one of Canada's rarest plants, can be found in the Thousand Islands.

duit between the much bigger landscapes to the north and south. In this way, the ecology of the Boreal forests of the Shield has a connection across the same granite bedrock to the forests of the Appalachian region. Plants and animals of both of these very different regions mix in the Thousand Islands. This is a very unusual ecological situation, a mix of vegetation and wildlife from the Atlantic coastal forests to the east, the Appalachians to the south, the northern Boreal forest, and the forests of the Great Lakes basin to the west. A number of species from these various neighboring forest regions near or at the limits of their normal ranges can also be found here. At the same time, the chemistry of the soils and the topography on the Frontenac Arch are, however, quite different from that of the sedimentary lowlands on the east and west flanks of the Arch. Because of this, east–west plant and animal migrations are inhibited, but not entirely prevented.

Plant and animal life from the forest region along the Atlantic coast has found its way inland along the St. Lawrence River valley. Similarly, elements of the Great Lakes forests, and even to some degree from the Carolinian and Prairie regions, have moved eastward through the basin of the lakes. Because of the river and shorelands, the potential barrier of the Frontenac Arch has been partially overcome. In this way, the Thousand Islands are able to host some of the elements of these far-reaching life zones as well.

The Thousand Islands boasts a wide variety and unusual combinations of plant and animal life. There are a number of factors: the landform links to neighboring forest regions, the tremendous number of habitat types created by so many variations in this rugged topography; diverse rock and soil types, each with different chemistry and water-holding ability; and a climate that is quite moderate for its latitude because of the proximity to the Great Lakes. In the nearly 10,000 years since the end of the last ice age, wildlife and plant life has re-colonized the region by way of the many routes provided by landform features such as the river corridor, the granite ridges of the Frontenac Axis, and the lowland plains to the east and west of the Axis. All in all, this is a very unusual geological and ecological situation, creating one of the richest natural environments in the world, especially at these latitudes.

Channel north of Hill Island. No two islands are the same in terms of their size, shape, landscape, or natural communities.

LAND HABITATS

No two islands in the Thousand Islands are the same in terms of their size, shape, landscape, or natural communities. There is, however, an identifiable character and pattern in the lifescape of the Thousand Islands. Take a walk down the length of an island.

At the west end, the predominant southwest winds sweep over shores of tumbled and broken rock. Exposed to the hot afternoon sun and the drying wind, the coarse, sandy soil is very thin. Gnarled pitch pine, white oak, and juneberry trees cast only a little shade on the juniper bushes, blueberries, and sparse grasses that cling to the pockets moss and lichen-crusty earth. The air is resin-fragrant from the sun-heated vegetation. Dried leaves and branches rattle, not rustle, in the breeze. This weather-beaten, sun-burnt environment is the island windbreak that makes the communities slightly more inland and downwind a little less hostile.

Next, inland and upslope, is a forest of white pine, red oak, ironwood, and white ash, with a ground cover of sedges, trilliums, and wild sarsaparilla, to name just a very few species. There is always something new taking its seasonal turn in this forest under-story. Where at the island's west shore there was little wildlife, save for the song sparrows that dive for cover at about the first moment that they're spotted, this pine and oak forest is alive with scurrying meadow voles, gray squirrels, and, depending on the time of day and place, porcupines and raccoons, as well as brilliant orange and black orioles, yellow warblers, and downy woodpeckers. If this were a mainland habitat, one would expect to find chipmunks and deer mice scurrying about. However, these animals are found on only a few of the islands. Chipmunks seldom swim from place to place, and because they are deep in hibernation during the winter, they don't have the opportunity to cross the frozen river channels. Deer mice don't like to venture across open areas: they keep to the cover of their forest runways.

These upland slopes are quite bright and often sunny right to ground level. The forest floor is totally plant covered, unlike the ground in the more mature, canopy-shaded

woodlands that grow on the deeper soils of the valleys between ridges. This environment is just semi-dry, and root systems of trees are precarious, often spreading out in the thin soil over the bedrock. This is evidenced by the numbers of trunks of trees blown over in storms in years long past. The trunks are slow to rot and they repose on the stilts of their sturdy limbs. The flat-bottomed skeletons of their roots tilt vertical like the bases of tipped-over wine glasses on the table-smooth bedrock that lies just a surprising few centimeters below the soil. Heaps of broken stone collect where they dropped from the clutch of twisted roots. In life, the tree roots had tried hard to penetrate even the smallest cracks in the rock but could only provide a platform on which the tree balanced against the force of the wind. Here and there, mossy piles of stone hint at trees that have finally been recycled to the earth.

Basins of clay, remnants of lake-bottom sediments from the post glacial lakes, lie between the ridges of granite. These are found at the most sheltered heart of the islands. Because of the rich and deep soil, and because they are sheltered from the winds that sweep around the island shores, these forests have the most luxurious growth of the islands. At one time, in the centuries before this region was thoroughly lumbered for the timber trade to Europe, colossal oaks and beeches rose from these beds of clay. Since it has been nearly two hundred years since some of these island forests have been cut, there are second growth trees that have reached considerable size, though never in our lifetime will the immensity of the originals ever be seen.

These pockets of rich forest are quiet and still at ground level. The canopy overhead is nearly complete. Little direct sun reaches the forest floor. The earth is damp and soft, deep brown and pungent. It is laced with fine threads of rootlets of ferns, herb robert, maple seedlings, and hundreds of other plants that take their turn in the seasonal change of this

Grassy upland slopes on Hill Island support a wide variety of plant and animal life.

forest's composition. In spring, before the leaves unfold on the trees overhead, there are immense stands of white trilliums, dotted with patches of red trilliums, may apple, trout lily, columbine, barren grounds strawberry, early flowering saxifrage, and many more wild flowers of the deciduous and northern forests. Later, when the canopy overhead limits the amount of light that can reach the ground, the forest floor is a subtle greenery of currant bushes, doll's eyes, ferns, and tree seedlings.

Along the island's north shore and steep slopes of tumbled boulders flanking granite ridges are the hushed green stands of hemlocks. In these places, the land slopes to shores and canyons in such a way that soil can accumulate only where it is trapped between the broken rocks. Even after thousands of years, the soil depth is measured only in millimeters, but because the ground is shielded from the sun by the granite cliffs and ridges, the earth always remains damp and cool. One of the few trees that can take advantage of such an environment is the hemlock. Despite its fine, lacy leaves of needles, the foliage of the hemlock is so thick that it almost completely blocks the sun from reaching the ground. The dusky green of the hemlock needles and the rusty gray-brown of the bark almost seem to color the air itself. Very little other than stump-top mounds of avocado and lime green mosses, polypody ferns, colorful fall fungi, and striped maples grow in these shady places. The hemlock stands are so still that any sounds seem to come from outside, and yet to the observant, there are numerous signs of animal life. Red squirrels dash from trunk to bough, harvesting the hemlock cones. White-tailed deer find shelter here from both the heat of the summer and the fierce winds of winter. Chickadees flit in and out with what seems to be an unbounded energy. Blue-spotted salamanders stalk bugs in dark damp cavities under root-split stones and rotting logs.

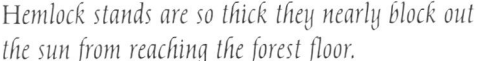

Hemlock stands are so thick they nearly block out the sun from reaching the forest floor.

Open, rounded ridge tops are a real contrast to either hemlock stands or deep forests. These are places where the bald rock is so smooth virtually no soil collects on its sloping surfaces. There is a constant tug of war between the lush growth of spring and the short but intense droughts of summer. In spring, a little growth of grasses and saxifrage begins on the thinnest of lichen-born, grainy soil. This meager vegetation withers and crisps in the baking heat of August, washing down-slope in the cold, heartless rains of November. In millenniums of seasons, the growth of would-be forests has crept only millimeters up these slopes. The ridge tops have succumbed gradually to the erosive powers of heating and cooling, freezing and thawing, and the prying of lichens and grass roots. The crystal grains of ridge-top rock have been nudged downhill by gravity and its ally runoff to contribute to the soils of the valleys and the sands and mud of the river bays. These are places where hemlocks peer up over the cliffs on one side and where white pines and oaks have been halted in their up-slope advance on the other by the lack of soil. Around the fringes of the ridge-tops are mounds of sedges and grasses, straggling staghorn sumacs, stunted white pines, the occasional pioneering pitch pine, and pillows of mosses that are alternately lush from rains or crunchy from the lack of it. The very tops of the ridges are the sole domain of rock-hugging lichens in many, many shades and hues and shapes that reveal their numerous species.

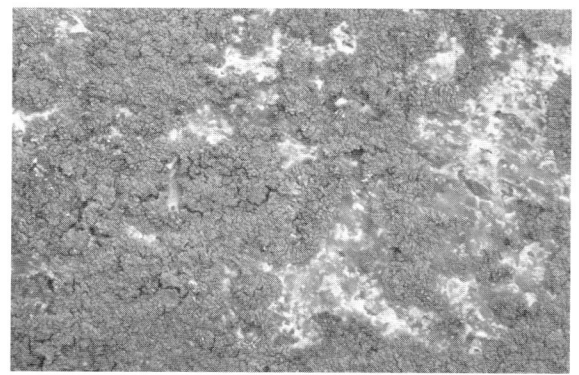

Lichens colonize even the driest and most exposed rock faces.

In east-end island bays and along sheltered shores, it is difficult to draw a line where the land and water interface begins and ends. Sediments of soil, eroded off the land or deposited in the shallows by the eddy and slow of river currents, collect in these bays and shorelines. There is a gentle merging of land and bay bottom, and because the water level varies by as much as a meter from summer to winter, the boundary line shifts seasonally. Waves of any size seldom reach such shores, leaving the interface fairly undisturbed. The stringy red roots of willows are awash in tangled mats, while the branches overhead lean out to touch their own reflections in the water. Algae-fringed stones are sometimes immersed, sometimes drying along a shoreline by birches, poplars, and the less-tasty alders that the beavers have somehow overlooked. Reeds hold tight to the sandy bottom but prepare to seal their own fate by collecting the decaying litter of leaves that settle around their stems, creating a niche that spadderdock and cattails will soon take the opportunity to fill. Ever so slowly, the shoreline will advance into the water, which becomes shallower with each passing decade because of the collection of silt and sediments from erosion and plant decay. Mink stalk along the shoreline to prey on the frogs, turtles, and snakes also foraging here. Green herons lean from willow and alder boughs, peering into the water, ready to spear at schools of minnows. Gadwalls and mallards tip for treasures of snails, waterfleas, and tender shoots of plants. Great blue herons oversee the tranquillity of these domains, digesting the scene along with their catch from the shallows. Many of the mammal members of the island community come here to drink.

AQUATIC HABITATS

Aquatic habitats are as varied and as patchwork as those on land. The same terrain of ridges, cliffs, and valleys exists underwater as on land.

Shoals are islets that never quite made it to the surface. They are, just like their above-water cousins, subject to the seasonal and daily forces of nature. They are heated by the sun, scoured by ice, pummeled by waves, abraded by current-borne sand, and split by the freezing and thawing of river ice. Shoals are immensely rich natural communities. The silt that gathers on their surfaces is full of nutrients from decaying plant and animal matter and so provides for a food chain that is long and complex. Plant and animal plankton abounds. Snails graze in the sediment and on algae, as do waterfleas and aquatic insect larvae. Clams and molluscs filter-feed in the rich water. Minnows hatched in the spawning areas around the shoals feed on the plentiful smaller animals and themselves become prey for successively larger fish, while snapping turtles and water snakes prowl for remains of these fights and for food they can catch on their own. There are plenty of refuges and hideaways here, given the number of crevices and tangles of plants as small as algae and as big as meter-long millefoils and pondweeds.

Out in the bays and on shallow flats, often well out from the island shores, are muck and silt-bottom environments where large beds of reeds and cattails grow. These seem to be fairly simple habitats, but this is not the case. Reed beds form in water that is usually a little deeper than where cattails are found. Waves sweep and currents sieve through here, carrying cargoes of silt and flotsam downstream. Some of this collects among the hardy stems, enriching the bottom sediments. The reed beds become hosts for other plants that may not have been able to get a foothold without some assistance. Pondweeds and millefoils clump amongst the reeds and create more shelter for pike, bass, and carp. These fish forage for prey that are in turn feeding their way down the food chain.

Cattail mats are more serene environments than the reed beds. Interlocking coarse mats of roots anchor to the bottom or often float up free if the water levels so dictate. In terms of biomass produced annually, this habitat is a leader in the region. An immense amount of plant matter has been produced over the years by the cattails. As the cattail stems and leaves decay in fall and spring, the litter gathers atop the mat of roots and becomes sites for other plants of the marsh to grow. Mints and coarse sedges are among the first to move in on the younger mats. Bur reeds, buttonwood, and dogwoods take hold as the amount of decayed plants builds up. Finally, the cattail marsh grows in so solid that it becomes difficult to discern whether it is actually a land or water environment. Roots of the shrubs that started out atop the mats have penetrated to the richly organic marsh bottom and have taken hold. Fish are no longer able to find refuge under the mats, and, where bays are big enough, the cattail colony has moved further out from the shores because the bay is filling in with plant matter. The increasingly swampy environment becomes more and more complex and it teems with life. Chorus frogs, leopard frogs, and water snakes spend much of their lives here, and are both prey and predator. A little further up the food chain are mink, herons, bitterns, and marsh hawks. Down the chain are countless dozens of insect species, of which the dragon fly is undoubtedly the most flamboyant member while the mosquito is the most notorious. Most habitats are divided up into territories, each important to its various inhabitants, but nowhere is territorialism so pronounced as in the cattail marsh, where the red-winged blackbird grips swaying stem tops and flashes its red patches to calls of 'chickoree' at other members of its species.

Aquatic environments: (clockwise from upper left) reed shallows, rocky shoals, willow-lined bays, and marshy bays.

MAMMAL RESIDENTS

Much as the Thousand Islands are stepping stones linking northern and southern natural regions of the continent, the islands are also linked on the local level to mainland natural communities. In a very real sense, the islands are extensions of natural environments on shore. If not for the occasional and random replenishment from neighboring islands and mainshore communities, the wonderful diversity of the island habitats would be lost in the long term due to failures of some species during hardships of extreme winters or drought, or from invasions of an overwhelming predator or disease. Occasionally, one can see the process in action: a branch, with seed pods or cones still attached drifting onto a shore; a squirrel, raccoon, or deer swimming a channel; or something so subtle as a tiny caterpillar drifting airborne on a silken thread. Thousands of mechanisms of dispersal are at work, year round — seed-carrying birds and winds, flying insects, animals scampering over river ice. The result is a natural environment in the Thousand Islands that is ever so dynamic. For the long-term ecological health of this region, sufficiently large areas must remain intact as living reserves for the area as a whole.

'Wildlife' has meant many things over the centuries. To First Nations peoples, wildlife was a source of food, particularly in the earliest days before agriculture evolved, but wildlife also became characters in many legends. For the most part, these small groups of natives just summered here and probably had little effect on the mix of wildlife species. When Europeans began to explore and settle in the Thousand Islands, though, things began to change quite dramatically. To them, wildlife represented many things — food for the table, income from hides and pelts, a threat to individuals and livestock alike, a menace to crops, and revenue from supplying game to local and city markets. It didn't take long for all things natural, from habitats to inhabitants, to undergo significant change.

Before the 19th century, the wild lands of this part of North America had a different mix of large mammals than seen today. Wolves, moose, lynx, martin, black bear, eastern cougar, fisher, and probably a few more mammals were members of the neighborhood. All of these were eventually lost to the region. Hunting by farmers and settlers certainly

Chipmunks dart for cover under roots and rocks. (Parks Canada)

had a role in this — there are numerous accounts of kills that would be seen as wildly excessive today — but the major reason for these larger mammals disappearing from the local scene was loss of habitat. Land cleared for farms and settlements and a growing network of roads meant that before long, there simply wasn't food and shelter to maintain a population of large animals.

Ironically, the same development of the landscape that eliminated the first species mix created the circumstance for another. Some of the common mammals of today's field and forest are most successful in younger growth forests, 'edge' habitats, where old fields and meadows abut woodlands and in the old fields themselves. For example, there were no cottontail rabbits here when the settlers arrived. A southern animal, the cottontail made its first appearance in the Kingston area in 1925 and has been abundant ever since in its favorite habitat of shrub lands and overgrown fields. The coyote, too, is a relative newcomer, having found its way here in the early 1900s after spreading along the north shore of Lake Ontario, where its favored food of mice were having a field day, so to speak, along the messy margins of crop and pasture land. White-tailed deer were likely always present here, but they, too, have benefited greatly from more edge and shrub lands, combined with the elimination of most of their predators.

All in all, there have been 37 species of mammals, including man, recorded in the Thousand Islands, and there are a dozen or so more with the potential of turning up as yet. Situations do change. For instance, several sightings of the fisher have been made recently, as has been the case with black bear, moose, and opossum. River otters, those playful critters that love to slide down snowy shoreline slopes, are becoming more and more common. Bats, mice, and shrews are understudied in the Thousand Islands, and new discoveries may yet be made.

UP IN THE AIR

Although the Thousand Islands is not one of the world renowned meccas to which bird watchers 'flock', it can certainly hold its own in terms of variety. This, once again, is due to the tremendous number of differing habitats. From hawks to shore birds, waterfowl to woodland species, there is plenty to see. The different species seen here over the years now numbers about 250. Viewing these species, even during spring concentrations, can be a challenge, though, since there are no points through which flights are funneled. The birds tend to use all of the islands as stepping stones when crossing the river, or may work their way along either shore.

Each season has its own special cast of characters. The spring specialty is waterfowl. In late March and early April, enormous rafts of ducks appear as if overnight, concentrated by the hundreds and thousands in relatively shallow and sheltered waters where spring breakup has opened channels and bays. Single day viewings of over 10,000 ducks are not uncommon. The Thousand Islands is a staging area, where greater and lesser scaup, canvasbacks, redheads, mergansers, golden eye, scoters, and many more gather to rest, feed, and begin their courtship rituals while awaiting that mysterious signal telling them the ice is leaving their northern summer nesting grounds. Overhead, there is a steady stream of V-shaped flocks of Canada geese, sometimes just pencil-thin lines in the distance, making their way noisily along the river corridor to northern bays and tundra. On sunny days, when the warming land raises heated columns of air, turkey vultures, occasionally 10 and 20 at a time, soar gracefully over the higher ridges. Many of these will nest here for the summer.

There is almost a particular sequence to the migration of spring birds. Winter begins to loosens its grip here in mid March. Around about St. Patrick's Day, red-wing blackbirds, robins, and great blue herons make an appearance. Goldfinches, grackles, and the summering population of gulls are not far behind. By early April, mornings are all the more cheerful with the song and flights of many new arrivals, including phoebes, meadowlarks, tree swallows, ruby and golden-crowned kinglets, house and winter wrens, sapsuckers, flickers, cedar waxwings, a host of sparrows, and so many more. It seems by the time that the northern orioles, rose-breasted grosbeaks, catbirds, thrashers, warblers, and tanagers arrive, there could be no place left to perch and no more room for song.

Winter is considerably more quiet, but it is not without its characters. Blue jays, white-breasted nuthatches, dark-eyed juncos, chickadees, and evening grosbeaks are among the ever-present. Black-backed gulls come inland from the coast to winter, great horned owls reveal themselves by February as they hoot from nesting areas, and

The statuesque pose and 'qworking' call of the great blue heron are a common sight and sound in the Thousand Islands.
(Parks Canada)

common mergansers crowd into waters kept open by swift currents. The most spectacular of the winter bird residents is the bald eagle, which, like the merganser, concentrates around swift waters to catch fish — although it isn't beyond a little taste of duck or winter-killed deer now and again.

There are a few very large colonies of birds in the Thousand Islands. The great blue heron is probably the signature bird of the region, popular for both its outrageous language and oddball antics, as well as its grace in flight. While it roosts and nests in several places, the largest colony is at Ironsides Island, just off the edge of the main shipping channel east of Alexandria Bay. The din heard while drifting past the island in early summer is incredible. Other 'colonists' are ring-billed gulls and common terns. These raucous birds favor large barren shoals and islets along the river from one end of the region to the other. Common terns are becoming less common than in the past as they are not up to size when it comes to competition for territory on the nesting grounds with the gulls. Ring-billed gulls made quite a comeback in the last part of the 20th century. They were heavily hunted in the late 1800s for their feathers, which adorned the hats of society's finest women, and they were decimated by the effects of DDT in the mid 1900s. Both the gulls and terns are now in serious competition with another comeback kid, the double-crested cormorant. The cormorant was actually scarce on the river until the 1990s, when its numbers grew by leaps and bounds. Its erect posture as it stands on small shoals and its fast, straight-line flight formations are now familiar sights.

HERPTILES

The black rat snake is an endangered species in Canada. Painted turtle at Jones Creek.
(Parks Canada)

Without doubt, the most maligned and misunderstood group of animals on earth are reptiles and amphibians. They have been gifted through myth and legend with grotesque powers as symbols of doom, even seen as having the most cunning powers of persuasion. Most properly considered as valuable members of any natural community, reptiles and amphibians are known collectively to biologists as herptiles.

There are so far 29 species confirmed as residents of the Thousand Islands. This is indeed a significant number, and distinguishes this area as one of the most herptile rich in all of Canada. Among the five species of salamanders recorded, the mudpuppy is the rarest. It is a completely aquatic fellow who rummages along the river bottom on the hunt for aquatic insects. The most common is the red-backed salamander, often encountered under rocks and logs in damp places throughout many types of woodlands.

Nine frog species have been noted. Because they have such closely defined habitats, and there is such a variety here, frogs seem to turn up everywhere. Eastern gray treefrogs adroitly travel the tree tops, capturing insects, changing color to match the surface on which they sit, calling shrilly and loudly on early summer nights. The smallest frog here is the northern spring peeper, which tends to be found in larger numbers at suitable swampy fens and woodland ponds, and peeps with enormous volume, considering its size, on early spring nights. Colonies of this wee animal are becoming increasingly rare over all of its range, and it is considered by biologists as a sort of warning flag of trouble on the habitat front. The bullfrog is the largest of the frogs. Its 'jug o'rum' call seems universally known. Open marshes with lilypads or spadderdock are the hideaway of this frog, as it rumbles out its song all the summer day, and well into the evening.

The number of snakes in the Thousand Islands ties with that of the frogs, at nine. Again, this reflects the diversity of habitats found here. The black rat snake is the largest and among the rarest of the snakes in Canada, although it is not uncommon here. At up to 2.5 meters long, shiny jet black and gold flecked in color, this is a hard snake to overlook, especially since it often hangs out around barns, open spaces, and roadsides. The black rat snake is also an accomplished climber, where it works its way into treetops, searching out nests for young birds and eggs. On the ground, it forages for frogs, rodents, and perhaps, as the name would suggest, rats. During the summer, this snake will range over distances of several kilometers, but in fall, returns to a communal den, or hibernaculum, deep in a hillside of broken rock. On occasion, people walking a shoreline stumble across what they at first think, because of its size, is a black rat snake. Some, knowing that the rat snake has a reputation for a mild mannered temperament, will pause while trying to make this other snake's acquaintance when it threatens to strike. The only thing that the two snakes have in common is potential girth. While both can reach sizeable diameters, the northern water snake is darker brown and black, never as long as the black rat snake, and short on temper. If cornered, the water snake will strike out and then flee to the water, where it dives and swims very well. Neither of these snakes, nor any other in this region, is poisonous. The smallest snake here is the northern brown, so tiny and mottled that it generally goes unnoticed as it forays among grass and twigs for earthworms and sow bugs.

Of the five turtles, the largest and most recognizable is the snapping turtle. It is most frequently seen in early summer, when it hauls itself, usually dripping with strands of algae, out of its favorite marsh and heads off to nest. With so many county roads criss-crossing marshlands, the gravel beds that the snappers desire often bring them into the path of cars. Another turtle, generally smaller in size but rough-backed like the snapping turtle, is the map turtle. The map turtle gets its name for the ridge-like lines on its rugged shell that look like the elevation lines on a topographic map. This turtle suns itself on rocks along deeper river channels, where it dives to feed on snails and crayfish. Painted turtles are the most common, usually seen lined up on logs or marsh edges in sheltered bays and ponds. It would be a toss-up to say whether the Blandings turtle or the stinkpot is the rarest here. Both are uncommon in Canada in any event.

Last, but not least, on the herptile list for the Thousand Islands are the five-lined skink and the American toad. The skink, a very small lizard, is quite secretive, dwelling in hollow logs where it darts after insects and worms. It is rare in Canada. The American toad, though, is very common on the mainland. Because it is not generally a swimmer, it is found on only a few islands. It loiters in damp, shady places on the lookout for garter snakes that would consider it dinner, while it watches for insects that it considers fair game for itself.

IN THE DEEP

A book about the Thousand Islands couldn't be complete without a discussion of fish. The landscape below the water's surface is every bit as complex as above. Gullies, cliffs, open and shallow flats, boulder-strewn slopes — it's just like above water, but without trees. There are many different habitats here, and so there are a niches to be filled by a great many species of fish, aquatic insects, snails, invertebrates, crustaceans, and shellfish. A complex ecology developed in the watery environment, just as was the case with terrestrial communities. Populations came from the lakes and rivers to the east, west, and south. Those populations continue to be changed today, with many foreign elements added to the system, arriving by ship from the waters of overseas ports.

While there are no accurate counts for the number of species of fish present, a fair estimate would be 85 to 90. A few of these are among the most prized of North America's sport fishes, and have lured fishermen to the Thousand Islands for well over a century. The muskellunge, after the sturgeon, is this continent's largest freshwater fish. Records show the 'muskie', as it's nicknamed, to exceed two meters in length and weights nearing 45 kilograms. While a good catch these days would be a fish of half that size, there is always the fascination that the 'big one' may still be out there. Muskellunge spawn in weedy shoreline shallows where the young spend their first years trying to survive the host of predators, including diving beetles, sunfish, pike, and, as they grow, terns and herons. As they grow larger, they inhabit marshy bays and eventually, when reaching much larger size, move into deeper water. The muskellunge lurks around sunken logs or heaps of boulders, waiting for prey to pass through its neighborhood. It darts out after its fish victim, catching it sideways, and brings it back to the lair to dine. Ducks, muskrats, or any other animal that makes the fatal error of swimming in its territory are not beyond being placed on the menu. Other favored sport fish include large and smallmouth bass, northern pike, and walleye.

A successful method of fishing in the Thousand Islands is drifting with the current along the shore. This could bring an array of results, including a fair share of surprise snags, since the lure will pass over such variety of submerged terrain. But one need not be a fisherman to get in on the action. A huge selection of fish can easily be seen from dock or shore. Rock bass, sunfish, and perch cruise near shore, seeking patches of sun-warmed water in early summer and shady overhangs or the shade of moored boats when the water warms up. Schools of minnows, especially spot-tail shiners and bluntnose minnows, pursue their plankton and tiny invertebrate prey along shallow bays and shores, ignoring the bottom-feeding madtoms but watchful of darters, sticklebacks, and the veritable host

Rock bass, with their bright red eyes, are common in rocky shallows.

of other predators. Since the river is now so clear, mask and fins can make fish watching as enjoyable as bird watching. Drifting quietly along ledges and along the edges of weedbeds, one can come very close to an amazing variety of river residents.

When First Nations peoples came to the Thousand Islands to hunt and fish, and when settlement first began, fish in the river and tributary streams were so abundant that, as one account recalls, you could dip them out with a frying pan. There indeed was a reliable source of food! By the end of the 1800s, attitudes on fish were to change greatly. Industrialization led to enough leisure time that tourists could visit remote and bountiful regions such as this to fish for sport. Over the next century, fishing itself became an industry, which was not without effect on fish populations.

Just as there was a change in the makeup of mammal species here following settlement, so, too, the situation changed underwater. Introductions of water plants from Europe, such as the bay-clogging millefoil and curly pondweed, have greatly changed the light regimes, current flow, and nutrient levels in shallows throughout the river. Some species benefited for having more protection and shelter; others lost their clear and open spaces. In recent years, the zebra mussel has invaded. Its population has boomed to such enormous numbers that it has filtered the sediments and plankton from the nutrient-laden Great Lakes. The water is now incredibly clear, and light penetrates to great depths, allowing some plants to grow in much deeper water. A light-sensitive species, the walleye, now has less refuge — but this, too, was the case in centuries past, before nutrients from cities and farm runoff made the once-clear lakes so murky.

AND NOW, FOR THE WEATHER

To quote Mark Twain loosely, "Everyone talks about the weather, but nobody does anything about it." Indeed, weather in the Thousand Islands is so variable that it features in many a conversation. Another appropriate phrase here would be, "If you don't like the weather now, stick around five minutes." Forecasters face quite a challenge in making their predictions for this part of the continent. Systems that develop far from here are often in collision over the lower Great Lakes and St. Lawrence River valley. High and low pressure zones are pushed and pulled by systems that may have formed off the Pacific coast, polar regions, in the Mississippi Valley, the Gulf of Mexico, or on the Atlantic seaboard. Most of the weather systems originating in North America exit down the St. Lawrence River valley. Figuring all that out is merely to add an element of science to good old-fashioned guesswork.

If there is any steadying influence on the day-to-day weather, it is the Great Lakes. This enormous body of water is a heat sink; that is, it stores cold from winter and heat from summer, lessening the effect of temperature extremes throughout the year. Land in the Thousand Islands warms more slowly than areas inland in spring and cools more slowly in fall. The net effect is to lengthen the growing season here. By way of illustration, the majority of the islands have 160 frost-free days in an average year, but that number drops to 145 just east of Brockville, at the region's east edge. This moderated climate is a significant reason for the number of southern species of plants found here.

The rugged topography of the Thousand Islands changes climate patterns greatly over short distances. This microclimatic effect can create shelter or exposure, heated or cooled areas, and dry or damp places in any number of situations and combinations throughout he islands. Such an effect broadens even further the habitat opportunities for plants and animals, and lends to the diversity of life in the Thousand Islands

And now, for a few weather statistics. Summer temperatures average warmest in July, with the day and night mean at about 21°C. Daytime highs will occasionally exceed 30°C. On the flip side of the calendar, January is the coldest month, with day and night means at –8°C, although temperatures of –30°C can be reached. The sunniest period of the year is in July and August, when it is partly cloudy 64% of the time, but completely overcast only 14% of those months. Conversely, it is cloudiest in November and December, being overcast 52% of the time and just 15% of those months have clear skies. May is the rainiest month, averaging 11 days with measurable amounts. That frequency drops to an average of nine days per month from June to September. The frequency

Ships ice bound at Clayton in the 19th century. (Antique Boat Museum)

of precipitation is lowest in January and February. Snowfalls are seldom as heavy in the Thousand Islands as just to the south, along the east shore of Lake Ontario, where soggy winds off the lake rise, cool, and dump great amounts of the white stuff on the Tug Hill Plateau. That same westerly wind carries that same moist air down the St. Lawrence valley with much less snowfall effect. Although thunderstorms are not particularly common here, they are most seen in July, with an average of six to seven a year, but virtually none in winter. Winds blow most of the year from the west, especially the south west, as seen by the way white pine trees are 'flagged', or wind pruned, away from the usual wind direction. Southwest winds average the highest speeds, at 25 to 35 kilometers per hour. Winds speeds are highest in November, but average some eight to ten kilometers per hour more in summer than in winter. During the course of the year, winds can be expected from any direction, influenced by the broad spectrum of weather patterns seen in this area.

Water temperatures are tied, of course, to air temperatures, but highs and lows will lag a little behind because of the enormous amount of water that is annually heated and cooled. By late July and through most of August, the river averages about 20°C degrees. Ice begins to form in bays and quiet sections of the river in mid December, and it generally isn't until early April that these same places begin to thaw free. Ice cover is inconsistent from year to year, depending on the overall severity of the weather. Not all of

A thunderstorm on the eve of a hot summer's day.

the river freezes over. Fast currents in the area of the Thousand Islands International Bridge, in the Brockville Narrows, and in the main channel east of Grenadier Island will most often be ice free all winter long.

If the geology of the Thousand Islands shaped wildlife habitat, so geology set the stage for that other species of the region, man. Land and man. While the human perspective is to place man first in this phrase, we are no less determined by the lay of the land than the plants and other animals. The resources of the landscape decide whether man can actually put down roots and create a history. It's in this special setting — this unique living environment — that the history of man in the Thousand Islands is written.

3
History & Legends

A SUMMER PLACE

Sunsets over the waters of the Thousand Islands are truly 'world-class'.

The sun has dropped behind the silhouettes of the neighboring island and the glow of the cook fire at the campsite will soon be brighter than the evening sky. As day yields to night, the travelers shift their focus from the fading magenta sunset to the radiant warmth of the fire. The cooking fire has become the center of the campsite's universe. A longer stick pushed into the center of the fire. A stubborn log rolled into the red glowing coals. The sizzle and snapping shower of sparks as a dry pine knot bursts into flame. The light from the fire is so much brighter now, dancing with the coal-black shadows out amongst the trees. Tending the fire becomes a ritual of silent communication, a way of turning your back on the uncertainties of the night and kindling a companion among the flames. No words spoken; none needed. But the last wind of the day heaves a final sigh in the boughs of the tall white pines, and the air falls silent as if to listen to the tall tales that will inevitably be told around the fire . . . This campsite has hosted visitors for countless centuries. Some of the yarns told here have been of heroic deeds, others have been of simple pleasures, all part of the human history of the Thousand Islands.

The history of the Thousand Islands reads more like a campfire tale than the stuff of textbooks, a story created by people enroute to all corners of the continent and by others whose dreams were as simple as finding a living from the land. The story of summer hunting grounds and farming a rugged land, of traders in fur and lumber, of tourists and guides, recreation and conservation. The history of the Thousand Islands has been written by people touched by the events of the world, but distinctly flavored by the character of the land. They all have one common bond: a deep pleasure in the wonderful beauty of this landscape.

FIRST FOOTSTEPS IN THE GARDEN

While the history of the first peoples in the Thousand Islands spans thousands of years, the details of that story can only be read by those observant and sensitive enough to see the subtle clues not quite erased by nature. Erosion and decay over time have all but eliminated every trace of man's earliest presence. Even so, almost needle-in-the-haystack finds — a campfire and campsite, fragments of a dropped clay pot, a lost or discarded tool, or chips of unusual stone left where tools were made — suggest scenes of ancient life in the Islands. There are places to see where all of these things have been found: they may actually be within an arm's reach at a favorite lookout. There are places with even more dramatic features, such as paintings on rocks, but these sites are so delicate they are best left undisturbed.

The Thousand Islands has been a special summer place since the time when human history began here, some 9,000 years ago. Although shorelines around the Great Lakes changed considerably when the land gradually rebounded after the incredible weight of the glacial ice melted away, shorelines in the Thousand Islands are in pretty much the same place as they were thousands of years ago. There were only a few meters of variation in the water levels relative to the shore through all that time. The island scene today is very much as would have been seen by the first peoples, except, of course, that the character of the forest and river life has evolved and has been altered.

The first hint of human presence here is a primitive stone tool found on Gordon Island, dating from somewhere around 7,000 BC. The mixed woodlands of pines and deciduous trees were still sparse compared to those of today, and the summers remained cool. These people were nomadic, with small groups combining their efforts to hunt large prey and migratory herds. Such peoples probably came here very infrequently, traveling to places that would help them find materials needed to make hunting points and to track down seasonally moving prey. These ramblings would allow this small population — there were probably just a few hundred people in southern Ontario at the time — to exchange ideas, trade, and find new mates. A few thousand years would pass before the environment became rich and diverse enough to support larger populations on a more permanent basis.

Eastern North America is rich in waterways. Since these are the easiest routes to navigate and because the shore marshes are excellent habitats for prey, aboriginal populations grew up along them. Over the passage of several thousand years, forests, fens, and marshes became ever-richer, able to supply the growing human population. There was less need for people to move about in search of food. Trade networks developed, bringing copper nuggets and tools from the north and shells from the east. Woodworking tools became more complex as more trees were cut for campsite clearing and furnishings. Local cultural identities began to form and ideas of territory came into being.

When the forests finally flourished, the first peoples would have found the Thousand Islands ideal for summer life. The river provided a bounty of fish and other animals. The forests were alive with birds and small prey. River woodlands provided shelter from both the sun and from summer storms. Points and headlands were ideal lookouts — and breezes there helped keep mosquitoes away. In winter, life in the islands would have been another matter. With the leaves gone, there was much less shelter from the weather. Winds would have buffeted the islands from every direction and swept across the frozen river, hardly slowed by the trees on the smaller islands. Waterfowl would have migrated as winter sets in and fishing ended when the river froze. The hunting was better in the shelter of the mainland forests, where habitats were large enough to support the large animals that do not hibernate. The people moved inland where both shelter and hunting were considerably better.

Lost Channel and Rainbow Span of the
Thousand Islands International Bridge.

Tom Thumb Island (above) just barely gets island status with "two trees and six square yards of grass."

Storm waves bend around the Spectacle Islands near Howe Island (top right).

A classic pitch pine and rocky shore, Sugar Island (bottom right).

Glassy swells reflect the colors of early morning, near Madawaska Island.

Old abandoned hull in Molly's Gut (opposite, top left).

Quiet marshy bays are important nurseries for fish and birds (opposite bottom).

Thunderheads building over Grenadier Island (opposite right).

A dark, still, and damp hemlock stand (left).

Wind-bent white pine at Fitzsimmon's Mountain — as the first settlers must have seen it (top right).

A young tree grows on a stout branch of an old wind-thrown pine (bottom right).

Treefrogs trill loudly on early summer nights (top).

Trilliums blanket deciduous woodlands early in May on Mulcaster Island (bottom).

A red fox on the way to an evening meal.
(Parks Canada photo)

White-tail deer are plentiful in the Thousand Islands.

*Icy mists at Ivy Lea Village,
where the river seldom freezes.*

For the very observant, there is plenty of evidence of the people who spent their summers in the islands hundreds, even thousands, of years ago. Campsites favored then were the same as we choose today. Access, shelter, clear views, and proximity to good fishing and hunting decide the best sites. At those points and bays, rings of fire-cracked rock show where campfires have burned. Some may have been used on and off for a very long time; only traces of them may show now, as the falling litter of leaves and twigs has almost covered the stones. Still others may be found in the shallows just offshore, dating to when the water levels were slightly different than today. Bits of broken clay pottery, fragments of flint, or even points from arrows and spears show signs of these camps.

A CONTESTED BORDER

The Thousand Islands has almost always been a border region — the river is a natural division between cultures and nations — and as a result often the site of conflict among nations. While there is no written record, only oral tradition (however roughly interpreted by the first European historians), to tell the story of the First Nations people, it would seem that in the few centuries before European explorers entered the picture, there were two main cultural groups in the Thousand Islands region. The Algonquins were the dominant people, controlling all of the territory in the lower Great Lakes, upper St. Lawrence River, and the uplands of the Adirondacks and nearby Canadian Shield. They kept the Iroquois under their control and jurisdiction. The strength and size of the Iroquois population eventually grew, and this society of farmers turned on the powerful Algonquin hunters, driving them from the river valley. A deciding battle took place at the site of Clayton when the Iroquois overcame a walled Algonquin village at the mouth of French Creek. That place became known as Fallen Fort.

The French came to North America not long after the territory-defining wars between the Algonquin and Iroquois nations. By about 1600, the French began to exploit the natural resources of this 'new' land, particularly furs for the rich fashion markets of

Early French explorers and First Nations people among the Thousand Islands.
(*National Archives of Canada*)

Europe. Since the best source of furs was in the lands of the Algonquins and the Hurons to the west, the French formed their alliance with these nations. Such unions often come at a cost — and in this case, the Iroquois became the 'adopted' enemy of the French. Since the lands of the upper part of the St. Lawrence were in the territory of the Iroquois, it was in the best interests of the French to bypass the Thousand Islands region during their travels inland, and certainly not to settle there. Not pleased with the trading partnership between the French and the Algonquins, the Iroquois often raided the fur-laden canoes of their rivals along the river. For safety, the more difficult and longer route up the Ottawa River and across the French River to Georgian Bay had to be used.

The French explorer Samuel de Champlain is credited as the first white person to see the Thousand Islands, even if it were only a distant glimpse. Champlain had traveled the northern route but in 1608 he was coaxed by his guides into venturing south along rivers and streams to the place that someday would become Kingston. Champlain was urged to take part in a raid on an Iroquois camp on the south shore of Lake Ontario. Figuring that a raid on the Iroquois stronghold would settle things for France, he canoed along the east margin of the lake, where his guides pointed out to him that there were many islands in the river. Quite an adventurer, but not much of a tourist, Champlain's only remark about the Thousand Islands was, "This lake is 25 or 30 leagues in circumference, and in it are beautiful islands, and it is the place where the Iroquois enemy catch their fis (furs), which are very plentiful." Champlain was wounded by arrows in the leg and knee during the intense fighting with the Iroquois, and wasn't in any condition to make an excursion downriver to explore the islands on his trip back across the lake. Instead, he returned to the safer route well north of the Thousand Islands.

Champlain might not have won a battle, but he did ensure a war. The Iroquois were not about to forget that the French took sides with their enemies, the Hurons and the Algonquins. By the mid-1600s, the Iroquois found themselves allies with

Dutch traders and colonists, who were settling the New England areas and were not adverse to assisting in the plundering of the French trade routes by supplying the natives with guns. All of the native peoples were suffering losses from diseases introduced by the Europeans. However, the guns bartered to the Iroquois by traders, but not presented to the Algonquins and the Hurons by their partners out of what may have been a prudent lack of trust, were the Iroquois' critical advantage. By the time an uneasy peace was made in 1653, the native allies of the French were scattered well to the north and west of this region.

It wasn't until 1654 when Simon LeMoyne, a Jesuit priest, became the first European to see the Thousand Islands firsthand. The journey had only just become possible because of the peace with the Iroquois. LeMoyne had traveled 10 hard days by canoe upriver from Montreal. That July he wrote: "The 27th. We coast along the shores of the lake, everywhere confronted with towering rocks, now appalling, now pleasing to the eye. It is wonderful how large trees can find root among so many rocks." LeMoyne continued further inland to carry on a mission among the Hurons, building an alliance to French interests.

Count Frontenac in the Thousand Islands.
(National Archives of Canada C13325)

In 1665, King Louis XIV of France appointed Jean Talon as the Intendent of New France, and gave him orders to carry war to the Iroquois, "even to their firesides in order to totally exterminate them, having no guarantee in their words, for they violate their faith as often as they find the inhabitants of the Colony at their mercy." That September, troops slogged through the wilderness into Iroquois country, but after nearly two months of hard going with little success, the effort was abandoned. The natives learned that the French were now a powerful foe and fled, leaving their villages to be burned.

With all of this, the French learned that chasing off through the woods after an enemy who knew the wilderness far better than they was futile. That realization led to an alternative plan, to develop trade using forts as forwarding and trading posts, a far more sensible alternative. Remy de Courcelles, Governor of New France, and a man who had some first-hand experience in navigating the upper St. Lawrence, decided to establish two forts on Lake Ontario: one on the north shore and one on the south, with a small warship for communication and protection.

In the spring of 1771, after forwarding a letter to missionaries inland, instructing them to tell the Iroquois that his intent was one of peace, he set out upriver. Courcelles had a bateau specially built for the trip, with a nine-man crew rowing the cargo of provisions. Fifty-six men in 13 bark canoes escorted the bateau. The trip was considered impossible in those days; no one had ever taken such a convoy up the river. To the astonishment of the troops and Indians alike, the trip was made in just ten days. Courcelles stayed just long enough to point out the best site for a fort and then the same day set out back into the currents for the return to Montreal. The round trip had taken just 15 days, and proved to all that the French now had the ability to take a large force and supplies to the foot of Lake Ontario quite quickly.

FORT FRONTENAC

Count Frontenac was appointed Governor of New France in 1672. Like Courcelles, he grasped right away the importance of strongholds on Lake Ontario. Time was of the essence, since the English had stirred the Iroquois to negotiate trade with the Ottawa nation to the north. They, too, wanted to locate their trade at the east end of Lake Ontario, and knew that the French governor did not have the full backing, and therefore the resources, of France. The new world was to have been a source of income through trade, and was not considered worth the expenditure of large amounts of money for military ventures. Frontenac, however, had a stronger view of what that trade could mean.

He proved his resourcefulness with a very bold move. Frontenac ordered settlers and canoes from each settlement as well as many of his own officers to form a flotilla to ascend the St. Lawrence. Two bateaux were built for the trip, each painted with bright decorations, and mounted with a total of six small cannons. The flotilla of 120 canoes, the two bateaux, and 400 men was an unprecedented force on the upper river. Frontenac set out on a fast pace for the foot of the lake, intending to go to Quinte, along the north shore of Lake Ontario, where he had attempted to arrange a meeting with Iroquois of both shores of the lake. Frontenac was met at a campsite at the head of the last set of the river's rapids, however, by messengers sent down river by his envoy, Robert Cavalier, Sieur de La Salle. They relayed that the north shore meeting might not be the right diplomatic move, since it would seem to favor the Iroquois of that shore of the lake. Not wishing that to be the case, Frontenac pored over a map of the east end of Lake Ontario sketched by La Salle, trying find an alternative. His officers agreed that a place at the mouth of the Cataraqui River could be just as effective as both a meeting place and a fort.

Three days later, on the 12th of July, Frontenac arrived at the entrance to Lake Ontario and was guided into the bay mouth of the Cataraqui. He explored the bay and shores until that evening, declaring it to indeed be a very suitable place for a fort. "In fact, they [the Iroquois] did not deceive me," he wrote "for they led me to the pleasantest harbor that can be seen; it is more than three quarters of a league in depth; its bed is only mud and there is more than seven or eight feet of water in the shallowest places. The river which forms it has six or seven fathoms at it mouth and for the distance of nearly three leagues which it runs up into the land to a fall, it is of such a kind that quite big ships could easily enter it. A point situated at the entrance puts the harbor which is thereby formed so much under the shelter of all winds that boats could lie there almost without cables; and at the far end of it there are meadows half a league wide by almost three long where the grass is so good and so fine that there is none better in France." This is the harbor now crossed by the La Salle Causeway in Kingston. The site of the fort is at the west side of the Cataraqui River bay mouth, and archaeological work has uncovered the stone foundations, visible on the north side of the road, just

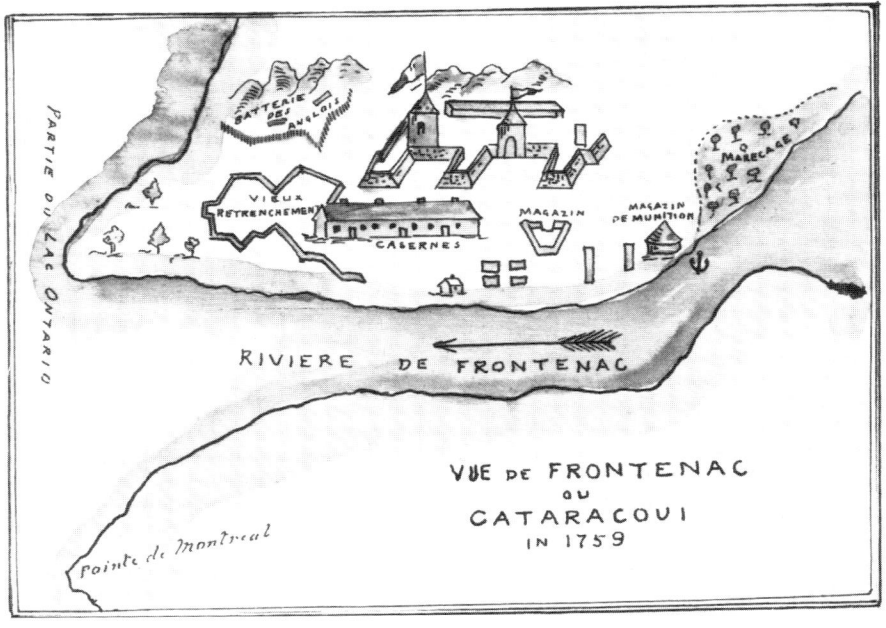

Fort Frontenac as sketched in the 18th century. Ruins of the fort have been preserved in Kingston.
(National Archives of Canada C6017)

where it turns from the causeway towards downtown.

Frontenac settled to work immediately, felling trees to clear the site. To impress the Iroquois of their resolve and strength, the French cleared the woods and roughed out a fort in just six days of hard work. Even while the construction was underway, Frontenac got down to the business of negotiating for trade and peace. With a rustic but elaborate spread of canvas sails on which to sit, and gifts of a ceremonial gun, tobacco, prunes, raisins, wine, brandy, and biscuits, the talks ranged from commerce to alliances. It took several days — and the talks nearly failed over a point about fixed prices for furs and trade goods — but peace prevailed.

Frontenac was proud to carry back the news of peace to Montreal, but he still did not get the support he hoped for from the governor. Instead, he was forced to finish the fort largely at his own expense. Nonetheless, his achievement meant that for the first time in decades, travelers and traders could voyage through the winding channels of the Thousand Islands without fear of ambush. The French had gained a strong grasp on this region, securing this part of the route to the source of the furs further inland. There was no colonization of the Thousand Islands by the French, however, because it was the will of the government that no energy or resources would be spent on settlements away from Quebec. This region remained a wilderness that saw only campsites on its shores and islands as traders took goods and furs up and down the river, and as the crews of supply boats travelled to the fort.

Because of the labyrinth of channels, the French voyageurs found a unique way to mark their passage on the river. They would plant a type of poplar tree, similar to the tall and slenderly compact Lombardy Poplar, at regular intervals and prominent locations on the shores. The distances between the plantings was called a "pipe" — the distance it would take for a well-packed pipe of tobacco to be smoked out. Since the poplar tree can sprout from a stump, even after an old tree has fallen, the descendants of the old tree markers survived for many decades, and some of those particular types of trees along the banks may trace their roots, so to speak, back to the original travelers' beacons.

A TALE OF A BROKEN PIPE

The weary crew of a small convoy of bateaux pulled ashore at Gordon Island one fall afternoon and set up camp for the night. They'd put their backs into the oars and against the wind for a full day, and it was time for rest. Tomorrow would see them arrive at Fort Frontenac and a roof over their heads, but this evening, like the last dozen, would be under canvas. The stiff breeze swept downriver out of the west, so they grounded the bateaux at the east end of the island in a little cove where the flat rocks step gently into the river. It was a good place for a campsite. They'd used it several times before. It was a place that gave a good view of the river and a spot that had good prospects for fishing and hunting as well. After all, there is nothing better than a bit of fresh meat to break the monotony of the tough, stale biscuits brought along as the steady part of their rations.

Duck looked promising this evening. They had seen plenty of them all day, gathering this season on the river before the weather turned cold enough to send them south. While they debated who would get out the gun and go for the hunt, pipes were brought out and smoked. After hard weeks of rowing the heavy cargo of tools and barrels up the current and against the wind, they were all a little tired and no one was quick to volunteer to chase ducks. The alternative of a bland meal helped out. Time to get at it. One of them lost footing on the wet rock where they'd trooped up from the shore, and fell. His pipe dropped and broke — there was a hardship — and the bag of fine but irregularly-sized lead shot spilled as well. He cursed and the others laughed, but the ducks eventually got hunted. The shot and broken pipe lay where they landed, just at the edge of the island campsite near the water, discovered by another camper over two hundred years later.

While this episode above is just speculation, the find of the pipe and shot near a campsite on the shore of Gordon Island in the Thousand Islands is real.

BRITISH CONQUEST

Over the next few years, Fort Frontenac was expanded, not by Frontenac but by La Salle, who had obtained loans to buy out the mortgage from Frontenac. La Salle tore down most of the palisades in 1675 and more than doubled the size of the fort. New buildings were enclosed in a defence of masonry walls on three sides that were up to a meter thick. A moat cut off the fort on the land side, a caution against attacks from the mainland. Farmland was cleared in the vicinity of the fort and some 40 cabins were put up in the area. All seemed to be going well in terms of both trade and the missionary work with the Indians.

All was not as well as it seemed. The Iroquois found the prices they were getting from the French at Montreal were not as good as those from the English at their colony along the eastern seaboard. The peace finally failed by the mid 1680s, and once again the upper St. Lawrence became treacherous to travel. When the governor of Canada, Denonville, mounted the river to Fort Frontenac with a force of 2,100 men in a fleet of 400 canoes and bateaux, the Iroquois took the threat for what it was and all final hopes for peace were lost. Denonville's forces fought the Iroquois at their villages in the Niagara region, and burned crops and camps there. In revenge, the Indians lay siege on Frontenac for a month and carried their retaliations down river to Quebec. As a result, the frontier was left to fend for itself, and only the sparest amount of supplies made their way to the lakefront fort. The Iroquois kept the fort surrounded, making attending crops, gathering firewood, and even getting water next to impossible. The situation was so hopeless that it seemed the fort must be abandoned. The commander of Fort Frontenac, Phillips Clement de Vuault, Sieur de Valrennes, was instructed to prepare to destroy the fort as he left. The walls of the fort were mined and would be lit from a long fuse. The cannon were to be oiled and hidden in the river, in case they could be retrieved in the future, and the wooden palisades were tarred to make them burn better. The barks, the small ships protecting the fort from dangers off the lake, were to be set afire. In any event, these could have been taken downstream no further than the head of the rapids, just below the present day site of Prescott and Ogdensburg. The bateaux were reinforced with planking along their bulwarks, and swivel guns from the fort were mounted on them. Valrennes expected to be ambushed by the Iroquois, whom he was sure would be waiting just down river behind Montreal Point, the present site of Fort Henry. The French had held out as long as they thought they could, but by early November, with the prospects of the hardships of winter looming ahead, the plans for the destruction were carried out.

Ironically enough, events unfolded that could have saved the fort. Three of the Indians who had been made hostages and sent to France by Denonville were released in a move to appease the Iroquois. As well, Frontenac himself had been reappointed as governor and he acted as fast as possible to save the fort bearing his name. He sent a convoy of 25 canoes with provisions up river, but realized he was too late when they met Valrennes at Lachine, en route to Montreal. In the following summer of 1690, Frontenac sent officers and men back to the fort to survey the damage. They found that the destruction of the fort had not been all that effective. All the wooden buildings had burned, but the stone walls were, for the most part, still intact. It would take six more years, but the fort was rebuilt.

In the meantime, war between France and Britain broke out, spreading from the continent to the Caribbean and North America. Both warring countries had limited resources to spend on the conflict in this region. It seemed that the scale of the war was dependent on the abilities of the English ally, the Iroquois. Since a smallpox epidemic from New England weakened the natives, the Iroquois sought peace, sending a delegation to Quebec City. Frontenac accepted. That peace was not long-lived, however, and fighting resumed after a few years when the Iroquois recovered sufficiently. The French, too, went on the offensive.

Despite his age — he was now 74 — Frontenac led a force of 3,000 men back to Lake Ontario in July 1696. After the fort was reprovisioned, the troops stormed across the lake to attack their enemy, wherever they were to be found. By the time a peace agreement between the English and French was signed at Ryswick in 1698, there was no doubt that the St. Lawrence and Lake Ontario could be firmly held by the French. Still, the prevailing opinion in France held that the efforts of colonization and commerce should be concentrated in Quebec, and so the regions of the upper river once again fell into a sort of no man's land that discouraged, if not actually prevented, settlement for at least the next 50 years.

By the 1750s, the French population of Canada was about 50,000. In comparison, the number of English in the Atlantic coast colonies had risen to nearly 1,000,000. The French were concentrated in a relatively poor farming and trading colony along the lower St. Lawrence. They were neglected by France and hardly able to improve their situation. The English were not any more considerate of their colonial subjects, but the British settlers were becoming less and less tolerant of their masters. At the same time, the English colonists were becoming more and more anxious to find new lands and opportunities, especially in re-routing the fur trade from French-held lands southward to New York.

It was pretty clear to the French that the English had ambitions in their territories, and although they wanted to take the offensive on the Great Lakes, their resources were tied up in skirmishes in Acadia and in Ohio. During the winter of 1755-56, a ten-gun schooner was built at Fort Frontenac, bringing the French fleet to a total of four ships. The fort itself, however, was showing its age. It was becoming apparent as well that its location was no longer the most ideal for proper defence of the region. Said one soldier, "When one of the guns on it is discharged, the whole fort shakes. Generally speaking, its situation is very bad, and it is of no use except for the stores that it would be desirable to protect against a sudden attack."

The spring of 1756 saw both sides trading attacks on the other. In August, Montcalm sallied across the lake and successfully assaulted the fort at Oswego, putting the French, for the moment, in control of Lake Ontario. English reaction was to blockade the mouth of the St. Lawrence, turning off supply from France and seriously weakening the French colony. The end for Fort Frontenac was approaching.

On the evening of July 25, 1758, Lieutenant-Colonel John Bradstreet landed 3,000 troops about a mile from Fort Frontenac. Some of his men tried to board the French ships in the fort's harbor, but were driven off. The following morning,

The British capture of Fort Frontenac.
(National Archives of Canada C2645)

cannons and mortars were dragged to the shelter of a small hill within 700 yards of the fort and opened fire. The English gained more ground over the next few days. Two of the French ships attempted to sail out of the mouth of the Cataraqui River, but ran aground in trying to avoid the bombardment of the English guns. The French were out-manned and outgunned; there was nothing to do but surrender.

Bradstreet allowed the 100 men, women, and children at the fort to leave for Montreal, on the condition of a promise that an equal number and rank of English pris-oners would be released. The English spent the remainder of the day of the surrender destroying the fort and its weapons, loading what provisions they could carry aboard their bateaux. The wilderness outpost that had stood at the entrance of the lake for nearly 75 years, but had never really had enthusiastic support from most of the governors of New France, was gone. Within two more years, New France itself had fallen to the English. Despite all of the years the French occupied the lands of the lower Great Lakes, today there is very little to show for it. There were no settlements, few place names, and only a few fort and outpost sites. The Thousand Islands remained a wilderness during the French tenure, seen only as a part of the corridor to the heart of the fur-rich continent.

UNITED EMPIRE LOYALISTS

Several more years would pass before any major British settlement would take place in the Thousand Islands. The plain truth of the matter was that, although all who passed by the islands marveled at the beauty of the region, they had no reason to stay. The farmland was not exceptional, seen as being too rocky and swampy for habi-tation. There were no apparent resources here for industry of any sort, nor was there any strategic need for a settlement. It took a major political and social revolution to drive settlers to the banks of the upper St. Lawrence and the Thousand Islands.

The British attitude towards its many colonies in the late 1700s was to rule, control, and expect a return for the protection offered by their flag — which was a little too much for the majority of the colonists along the Atlantic seaboard of North America, who, with the assistance of England's long-standing rival, France, wrested their independence from Britain. Opinion is seldom unanimous in a revolution, so it was natural that some citizens of the Thirteen Colonies would continue to be loyal to Britain. Some were reluctant to give up their allegiance to the Crown, others didn't find a place in the new order of things, and still others were obligated to leave because, in the minds of the

Fort Haldimand barracks on Carleton Island. (J.H. *Durham*)

revolutionaries, they had committed acts of treason. Many of these 'United Empire Loyalists' fled to Canada. Over 50,000 transplanted their roots north of the St. Lawrence River. After the war, they did not recover their land holdings in the United States of America.

With the turmoil of the revolution in their Atlantic colonies, Britain became concerned for the security of Canada. One weak point in the defence of Canada was the upper St. Lawrence River, where control of Lake Ontario and access to the river were an issue. Sir Frederick Haldimand, then Governor of Quebec, ordered a naval stronghold to be established in that area. Fort Frontenac, the old French fort at Cataraqui, was soon rejected as a site. Strategists saw weaknesses in its swampy surroundings and distance from the main channel along the south shore. The on-shore winds would limit ship maneuvers, and all of the suitable timber for shipbuilding had long since been cut. Instead, the site chosen was Buck Island, the large, almost round island between present day Cape Vincent and Wolfe Island. Its name was changed to Carleton Island, to honor a former Governor General of Lower Canada, Sir Guy Carleton, and the new fort was called Fort Haldimand. This fort was an impressive structure, with extensive ditches and earthworks laboriously blasted out of the rock, a blockhouse and barracks, and a shipyard. The shipyard produced large numbers of vessels for the naval effort. Most notable was the 22-gun brig-sloop, the *Ontario*, the largest sailing ship built on the Great Lakes to that time, which foundered in 1780 after delivering a raiding party to Oswego.

As the major British installation between Montreal and Niagara, Carleton Island also had a role to play in the settlement of Loyalists in this region. The Mohawks, an Iroquois nation that sided with the British during the American Revolution, had been driven from their lands and many had taken refuge on Carleton Island. When the treaty was signed in 1783 to end hostilities, the British stopped work at the fort, recognizing that this island close to the south shore would probably rest in American waters after new boundaries were sorted out. They relocated their defences to the old fort at Cataraqui and looked for land on the north shore to settle Loyalists. This land was in native territory, however, divided among the Algonquin nation — or Mississaugas, as they were often referred to by the British — to the west and the Iroquois nations to the east.

By this time, the British followed a policy of negotiating with First Nations peoples for rights to ownership of land. When brought by the British to Carleton Island for negotiations, the Mississaugas agreed to a treaty surrendering their rights — according to British correspondence, since no treaty document survives — to lands in the Thousand Islands and as far west as the Bay of Quinte. The British wanted agreements from the Iroquois nations, and so asked for similar rights in the lands on the north shore from the Quebec boundary west to the Gananoque River. Again, no treaty papers survive, but it was felt that lands had been secured for Loyalist settlement. As for Carleton Island, the outpost remained in British hands until the War of 1812, despite being awarded to the United States in Jay's Treaty in 1795 which settled boundary lines, though Fort Haldimand's armaments, shipyard, and even some of the buildings were moved to Cataraqui.

Macaulay House was built as a log house outside Fort Haldimand, then rafted to Cataraqui (Kingston) in 1783 and clapboarded.
(Heirs of Hon. W.F. Nickle, K.C.)

While past reports about quality of the land in the Thousand Islands had been less than glowing, the need to find places to settle the Loyalists inspired the surveyors to view the land in a new light. Perhaps the most influential appraisal of these lands came from Captain Justus Sherwood, a Loyalist who had experienced a number of run-ins with the revolutionary regime. "The climate here is mild and very good," he wrote in one letter, "and I think the Loyalists may be the happiest people in America by settling this Country from Long Sou to Bay Quinty." In words that would second Sherwood's opinion, the Surveyor General of Quebec, Mr Holland, wrote to Sir Frederick Haldimand, stating, "From the head of the Long Sault to the Top of the uppermost rapid where the Navigation begins the Country has a most favorable appearance, from hence to

Cataraqui the shore is high and Rocky, but opening here and there into beautiful Coves and Bays, where the view extends a great way into fine natural Meadows — and although the shore appears rough and uninviting the Soil is rich at some distance fit for all purposes of Agriculture, as I have been informed."

Despite all this encouragement, the lands first settled were to the east and west of the Thousand Islands, for reasons noted by Justus Sherwood's survey team. "We arrived at Carleton Island," he wrote. "There is a vast number of islands between Oswagatchie and this place, but in general, they appear to be barren rocks, excepting one called Granadier [*sic*] Island, which appears to be fine land." The first group of townships marked out for survey were to the east, from about Lake St Francis to the west edge Elizabethtown Township, which was the first township to be surveyed. Following Elizabethtown, the Front of Leeds and Lansdowne was laid out in 1788, Pittsburgh Township in 1789, and the Front of Yonge in 1794. Even after the surveys and allotments of land, the settlements were, for the most part, just lines on paper. For example, 15 years after the survey, the Front of Leeds and Lansdowne had a population of only 125 persons.

The townships were surveyed as blocks of land of about 10 square miles. A line was surveyed parallel to the river, and this was to be the first concession road. Lots that fell on the river side of the line were called the 'broken front'. The remainder of the concessions were placed at 1 1/4 mile intervals back from the first. The land between each concession was divided into lots of 200 acres each, and because of the spacing of the east-west roads and the size of the lots, the parcels of land were much narrower than they were wide. The purpose of this was to give as many people as possible river frontage, or at least the chance to be closer to the river. Sideroads, created by leaving spaces between lots every two to three miles, helped insure that there could be a way to get to the river from farms and villages inland.

The numbers of the lots were written on square pieces of paper and put in a hat. Ex-military officers drew first, from the lots along the river and on the best farmland, of course. Others would take the next best. Good farmland along or with access to the river was prized. Other favorable choices were well-drained and level lands with good soil back from the river. Some of the unlucky, however, would have drawn lots that were rocky or swampy. These were sometimes abandoned or perhaps sold or traded to other landowners. There are stories of whole lots bartered for a pair of boots, a calico dress, a meager amount of groceries, or farm tools. Some of the lots taken in trade this way were later sold for as much as two or three dollars an acre.

Many settlers in the Thousand Islands were soldiers and their families from disbanded regiments, such as Jessop's Rangers. Many more were individuals who fled to the region individually from the United States and Britain and were called 'unincorporated' Loyalists. There was a particular amount or lot of land that could be drawn by each man or family, determined by past military rank or circumstance. The first allotments of land were small, especially for those who had not served in the military. By 1788, grants were substantially increased, in hopes of encouraging more to take up life in the new land. A field officer could receive 5,000 acres, a captain 3,000, sergeants 500, corporals 400, and privates and non-combatants 200 acres. The Loyalist's wife and children were entitled to an equal amount of land.

The majority of these settlers came with few resources, financial or material, to begin their new lives. These were a people who had barely more than their resolve and ambition to make a new home in the wilderness, isolated and alone. Some settled into the task of clearing land for farms; others didn't stay long and passed up their opportunities for land here, taking their chances in other areas.

A crude log cabin built by early settlers of the Thousand Islands region.
(National Archives of Canada)

PIONEER LIFE

There is a somewhat romantic notion that life on the land for the early pioneers was a colorful and rewarding experience. No doubt, there were was an immense sense of achievement in making a go of it, but those were tough years, full of perilous hardships. Perhaps the two hardest years were 1787-1788, known as the Hungry Years. Summer droughts and long bitter winters brought on crop failures throughout this part of North America. British government distribution of seed and tools had also fallen behind. Crops that were planted either withered in the dry spring or suffered from early frosts. With only a few farms and villages scattered throughout the region, all just recently hewn from the wilderness, there was no means of supply from the outside world. Settlers were isolated. They had to take what they could from the woods: nuts and berries, buds from trees, fish speared in streams, birds and animals caught however possible. Bones were boiled again and again for soup, sometimes passed from farm to farm. Seed heads from grains that escaped the drought were harvested and boiled into gruel before they even ripened. Children took to begging food from boats passing on the river, if indeed those aboard had anything to spare. Farm lots were bartered away for a few pounds of flour. Some settlers died from starvation in the Hungry Summer.

Food was hardly scarce in the woods, though. Game was probably more abundant than in any other time in history, but these new settlers were not as savvy to the ways of wildlife and not nearly as knowledgeable as the native North Americans about edible plants. Many went hungry in the face of plenty. A son of one of the earliest and most prominent Loyalists, Adiel Sherwood, wrote in his memoirs about the abundance of wild animals. He reminisced about flocks of Passenger Pigeons so huge that they darkened the sky and could be knocked out of the air with fish-poles. A neighbor, he remembered, killed 30 with one shot. Sherwood also remembered flocks of ducks so large that when they took flight, the sound of their collective wing beats was as loud as thunder. Fish in some creeks could be scooped up with a long-handled frying pan. Deer and partridge were plentiful, and the woods and meadows were rich in strawberries, gooseberries, blackberries and cranberries.

As part of their settlement package, Loyalists were to have been given basic tools, including axes, augers, and hoes, seed for corn, wheat, peas, and potatoes, and livestock, such as cows, oxen, hogs, and chickens. The government's ability to supply and distribute, however, was not up to the demand, and some settlements, especially those distant from the source, got less than may have been their share. Some of those in the Thousand Islands fell into that 'have-not' situation and so the progress on farms was held back.

The government also issued grants to build mills near major settlements, but none were given in the Thousand Islands. The earliest mill to have been built near this region was at a falls on the Cataraqui River, just north of Kingston, but accessible from the river only by small boat. Farmers from as far east as Brockville would take their wheat to this mill to be ground into flour. In autumn, two canoes could be lashed together to make a stable enough vessel to take the precious cargo upriver and back. When the river was frozen, sleighs could haul the wheat and flour along the shore. The mills were a vital part of the farm community and would soon become an important part of the region's economy. Before long, mills were built by enterprising individuals at Gananoque, La Rue Creek, Brockville, and MacIntosh Mills north of Mallorytown. Mills closer at hand allowed more time to be spent on developing the farms. By the early 1800s, wheat production was so bountiful that this part of eastern Ontario supplied much of the flour that was shipped to the port of Montreal.

Some of the Loyalists granted land in the region had been prosperous landowners in the American colonies. When they fled to Canada, they were given very large amounts of land in compensation for the land they had lost and because of the roles they had played, politically or militarily. Gananoque's founders, Sir John Johnson and Colonel Joel Stone, were a case in point. Both recognized the value of the mill sites along the Gananoque River just north of where it flowed into the St. Lawrence. After some petitioning by both men, Stone was granted the lands on the west bank of the Gananoque River and Johnson the east, but while Stone drove the development of the town, Johnson did not settle in the Thousand Islands.

Water-power was the real key to the early success of Gananoque. The Gananoque River has several falls just back from the St. Lawrence. That power was not monopolized by any one industry but rather was shared by quite a number by letting the flow pass through races from one site to the next. Industries grew to supply the needs of the day, and towards the late decades of the 1800s, there was such an output from the town that Gananoque earned the nickname of the Birmingham of Canada — a comparison to England's most noted industrial area. There were flour mills and sawmills as well as factories producing a variety of iron goods, nails, hinges, forks, shovels, wheels and axles, and other farm machinery. Other factories turned out lumber, finished wood products, furniture, and, perhaps underscoring the burgeoning sportfishing industry, the Fluted Trolling Spoon factory manufactured fishing lures.

William Buell, who had fled from Connecticut with his family and then was active

The Paul farm near Brockville.
(National Archives of Canada C11709)

in the militia out of Montreal, was the first to build a log house on the present site of Brockville. He was soon joined in settlement by Adiel Sherwood, who opened the first tavern, Daniel Jones who built the first mill, and Charles Jones, who put up the first frame house. This group spurred on the development of the village, then called Elizabethtown, but all the while quarreled amongst themselves about what the name of the new settlement should be. The rivalry was apparently quite noticeable to outsiders, to the degree that the place was nicknamed 'Snarlingtown'. The story goes that Buell wanted to rename the village as 'Williamstown' and Jones thought that 'Charlestown' had a nice ring to it. Perhaps it was just odd coincidence that the proposed names were the same as the two men's first names. An early urban legend has it that Sir Isaac Brock, a very prominent military figure of the day, was asked to settle on the matter of the name while he was passing through town. After a moment of deliberation, he bestowed upon the town his own name, and so 'Brockville' came to be. Surrounded by good farm and development land, plenty of timber and streams, Brockville prospered. Industry grew, businesses expanded, and a jail and court houses were erected, reflecting the keen interest of some of the citizenry for the process of law. Brockville was one of the first towns to be incorporated in Canada, and has Ontario's oldest continuing daily newspaper. Nearby, in the village of Mallorytown, were among the first of Canada's glassworks and brickworks. In Lyndhurst, then called Furnace Falls, one of the earliest iron foundries was established, though the industry was short-lived because of the poor quality of ore available.

THE NEW FACE OF THE LAND

In a letter to historian William Canniff, Adiel Sherwood recalled stories told to him by his father, a man acknowledged to be the first settler in at least the eastern section of the Thousand Islands: "At that time, the country was a howling wilderness. Not a single tree had been cut by an actual settler, from the Province line to Kingston, a distance of one hundred and fifty miles." Within a few decades, a considerable change was made in the character of this "howling wilderness," as Sheriff Adiel Sherwood put it. Once-isolated farms were more substantial. The settlers' five-by-six meter log cabins with blankets for doors and mud-covered green wood chimneys standing at the edge of tiny stump-ridden clearings were abandoned once larger frame and sometimes brick homes were built. The lean-to buildings that once were the only shelter for livestock were replaced by log or timber barns. Sheds, smokehouses, woodsheds, root cellars, and ice houses were added to the successful farms, a tribute to the long, hard efforts of the settlers. Stone and brick houses stood along gravel roads and proud farmers looked out over rail-fenced fields and sturdy barns. Crops were reliable, and surpluses of grain, vegetables, and livestock built an income for the farmers. Niceties as well as necessities were available at general stores in the growing villages. Lumber mills, blacksmith shops, flour mills, and even some manufacturing of tools and carriages lent to the prosperity of the communities.

While man prospered, the natural environment of the Thousand Islands suffered from the development. Habitat for most of the largest animals of the region fell before the axe and fires of clearing. Moose, bear, timber wolf, marten, and fisher are a few of the animals that need considerable space and food to survive. They were over-hunted and trapped out. There was no longer sufficient habitat and suitable environment left for them. Other animals that took advantage of the clearing, though. The cottontail rabbit wasn't found in this region — it's actually a southern species — until its favored habitat of open and shrubby meadows developed on marginal farmland. A similar habitat benefits the white-tailed deer, and since there was more of this habitat and fewer large natural predators, their population boomed. Red foxes found more prey in the form of meadow mice, field-nesting birds, rabbits, and grasshoppers. Raccoons shared in the farm harvests, albeit not with the blessing of the farmer.

Smoke from burning brush and forests darkened and choked the air during the first years of settlement. Timber cutters selected the biggest and best trees, especially white pine, oaks, and hickory. When these were felled, clearings of considerable size were left. Bright sunlight reached the ground, in many places for the very first time in

*Old growth white pines — sizes never to be seen
again in our lifetimes.*

thousands of years, and let sun-loving plants thrive, changing the forest character. A growing number of the species encroaching on the wild were imports from Europe that had come to North America in discarded ships' ballast and mixed in with seed grains. Field edges and fence rows became a new type of habitat, with a low cover of shrubs and weedy species. The clearings, shrubby cover, and fields afforded food and shelter for birds, insects, and other animals that had only limited amounts of habitat in the past.

Furs may have been the most glamorous trade goods from the colonies, but it was the trade in raw materials, especially timber, that was the most vital to the British Empire. By the 1800s, trees of any significant size had pretty well been lumbered out of Europe. Large oak and softwood trees, needed for the hulls and spars of commercial and naval ships, had been brought down from the Baltic regions, but that supply became difficult to depend upon because both Britain and France, at war by this time, realized the strategic importance of the timber and blockaded the region. Timber meant ships and ships meant power. There was, however, an apparently inexhaustible amount of timber in North America. About the time that the dependable source was threatened at home, ships capable of carrying such heavy and bulky cargoes were able to make regular ocean crossings.

The demand for the lumber from the colonies seemed to be endless. Lumber merchants took wood from wherever they could find it, but of course, the closer it was to ocean ports, the lower their costs would be. East Coast forests were the first to be depleted. The next source was inland North America — and the St. Lawrence River was the access.

Even as the Loyalist settlements were being hewn from the wilderness, timber cutters were at work, often with disregard to the ownership of the land. For the next century, timber rafts were part of the regular scene on the river. The islands were the first to be timbered. After the suitable island trees were gone, the cutters worked further inland, floating logs down the

Gananoque, Cataraqui, Oswagatchie, and other rivers on the surge of the spring runoff. Cutting was selective because the market demanded big, clear, and straight timber. The real clear-cutting, for fuelwood for steamers, wouldn't come until decades later. Even so, the biggest trees would be gone within a few years. Oak and pine, one to two meters across at the base, once towered far overhead in the forests of the Thousand Islands, as dramatic as West Coast forests today — something never to be seen again in this region.

THE GREAT MACOMB PURCHASE

Following the War of Independence, events unfolded differently on the American side of the river. In New York State, the government duly noted the Loyalist settlement on the Canadian side of the St. Lawrence. Wanting to keep pace, the state created ten 100 square-mile townships along the eastern side of the Thousand Islands. The thought was to sell these to individuals who would take it upon themselves to carry the development costs. Before any of that land was sold, a man named Alexander Macomb, who had made a fortune in the fur trade, and his two partners purchased three and a half million acres on speculation. This included not only the original 'Ten Towns', but most of the remainder of the lands and islands in the upper river as far west as Oswego. This huge block of land, the 'Great Macomb Purchase', was bought for eight cents an acre — a bargain-basement price, even at that time. Shortly afterwards, Macomb fell victim to one of the many economic downturns that have plagued investors throughout history, and in his bankruptcy, the lands reverted to the state. Again the lands were up for sale, a task given to one of Macomb's former partners in the Purchase, William Constable.

Constable had a brilliant idea. He offered the sale of some of the lands to wealthy families in France, with the thought that they may have wanted to take refuge in a new land free of the persecutions of the French Revolution. One of those interested was Jacques Le Ray de Chaumont, to whom the American Government owed a considerable amount of money for his part in financing activities during the American Revolution. Le

Ray's son, James, had taken up residence in New York and had married an American woman. James had also managed to recover at least part of the debt owed to his father and had developed an interest in land speculation. Le Ray's brother-in-law, Paul Chassanis, also met Constable and bought more than 200,000 acres of land in the area of Watertown and the Black River in New York. Chassanis' intent was to clear land and build roads to attract French citizens to this haven in North America, a place he called Castorland. 'Castor' is the French word for beaver, a reference to the fur trade days gone by that would surely play upon the memories of the glory of New France before the British conquest. For his part, James Le Ray, now an American citizen, had been given custody of his father's land to avoid it being lost to the Revolution in France.

Le Ray took an active interest in seeing the lands in the Thousand Islands developed. His own mansion was built on the current site of Fort Drum, New York, and from that regal residence he saw to the establishment of other towns. These he named for his children and so became the communities of Alexandria Bay, Theresa, and Cape Vincent. He, too, encouraged emigrants from France. Joseph Bonaparte, the brother of Napoleon, and several other military men and their families were among those who came to settle the region, lending French names and flair to this northern corner of the United States.

A GHOST STORY

William 'Billa' Larue came to the Thousand Islands as one of the many settlers loyal to the British crown. He settled on the banks of a creek that tumbled in a series of falls into the St. Lawrence, about five kilometers west of the landing that served as the port for the village of Mallorytown. While some settlements closer to the seat of government were given grants to build mills, none were bestowed upon this region. Larue bent his back to the work of damming the low-lying area above the series of falls to hold back the water from the rains and spring runoff that would provide a steady source of power for his mills. Larue's income depended largely on the output of the flour and grist milling; the remainder of his land was too hilly and rocky for much more than gardening. He did have, however, a number of nut trees on the grounds near his big, two-story log house and barn.

The Larue mills were an important part of the community. Farmers would otherwise have to take their grains a considerable distance to have the grinding done. During the War of 1812, the production of the flour mill went to the military for a time, and it was considered necessary to protect the site. Rifle pits were manned where Larue Creek's last set of rapids froth into the river.

Billa Larue's enterprise netted him a considerable amount of money over the years. Apparently, he kept his savings in the form of gold, but hid it away even from his family. On his deathbed, Larue's wife pleaded with him to reveal where the money was, but he would not. Instead, he gazed out the window, across the grounds to where his children who had died some years earlier were buried, and proclaimed that was where his treasure lay.

Some people, probably Mrs Larue included, couldn't decide whether the treasure was buried on those grounds in a literal or figurative sense, but took up their shovels in efforts to find out. A lot of earth was turned, but nothing was ever found. One evening, however, a group of locals determined to get to the bottom of the matter. An account of the event is recorded in *The History of Leeds and Grenville*, published in 1879 by Thadeus Leavitt, then editor of the *Brockville Recorder*. The wording is so delightful that it would be a shame not to recount it verbatim.

"On a bright moonlight night, I, in company with three other men, left the Village of Mallorytown and proceeded to the vicinity of the old Larue mill, near the upper dam. We had provided ourselves with a witch-hazel divining rod, a goodly supply of shovels and picks; in fact, all that was necessary for an enterprise of such a character. All were in the best of spirits, and as the night was charming, we proceeded to the vicinity of the house, where Billa had resided, determined, if possible, to probe the secret to the bottom. We were under the guidance of an elderly gentleman, who claimed to be an expert in such matters, and had carefully instructed all engaged as to their duties. One command was imperative, viz: that from the moment the spot was indicated by the divining rod, not a word was to be spoken, happen what might. A short distance west of the house is the family cemetery, and in that direction we cautiously proceeded. The moon shone clear and bright through the pines on the overhanging cliff. Suddenly our director paused, the witch-hazel turned slowly in the direction of mother-earth. Retiring a few paces, our

Grave site at Larue farm.

leader re-adjusted the rod and moved forward, with precisely the same result. Evidently the secret had been solved and we were about to become the happy possessors of the long sought gold. Striking a circle, having a radius of about twelve feet, we removed our coats and proceeded to dig. How long we continued I know not, so intent were we upon our task. Gradually the sky became overcast with clouds, one by one the stars faded away, the moon disappeared in the vault of night, the wind sighed mournfully through the pines, yet not a word was spoken; darkness came down upon us like a great pall, our nearest co-laborer was only a spectre in the midnight gloom. Then came a rush of the blast through the overhanging trees, the blast was of icy coldness and penetrated the very marrow of our bones, though our bodies were bathed with sweat from our almost superhuman exertions. There was a trampling upon the earth in the distance, as if the guardian spirit of the treasure trove was marshalling all his cohorts to hurl back the audacious invaders who had thus dared to desecrate his domains and snatch away the glittering coin confided to his care. The excavation we had made was conical in shape, the centre being at lowest point, when suddenly there rang out clear and distinct in the night air, a sound which proclaimed that the pick had struck a metallic substance. A few shovels full of earth were thrown off, when with our hands we felt that we had struck upon what appeared to be a smooth flat stone or piece of metal; we have always believed that it was metal from the ringing sound which it gave forth.

"Redoubling our exertions, we removed the earth at one side, where we inserted a crow-bar, the point below resting upon some substance, which formed an excellent fulcrum, and which we concluded was the box containing the coveted treasure. With our united strength we slowly raised the covering, when in an instant we were surrounded by innumerable creatures, trampling up to the very edge of the circle. We could but indistinctly distinguish the forms of the new comers, but to my mind they appeared to be *black cattle*, and judging from the trampling, their number must have been thousands. We hesitated — a great fear came upon us, which I cannot describe — and, with a single impulse, we dropped the crow-bar, and ran for dear life. Beyond the house we came out of the ravine, near the new mill, where we paused. The moon was sailing majestically through an unclouded sky; the stars shone as brightly as when we first entered upon our task. We paused and consulted, and at last concluded that imagination had got the better of our senses, and that we would return to our work. This we

The Larue farmhouse, one of the oldest in the region.

did. We found the excavation, the coats lying on the ground, the crow-bar, shovels and pick-axes, but not a sign of the flat stone or metallic covering at the bottom of the pit which we had dug. Our leader sorrowfully shook his head, and declared that future efforts would be of no avail, as the treasure *had moved*. We gathered our implements, and departed for Mallorytown, fully resolved that in the future other searchers were quite welcome to secure the hidden gold left by Billa Larue."

WAR OF 1812 – 1814

The War of 1812 certainly wasn't the brainchild of anyone in the Thousand Islands. Most of the people here in the early 1800s could trace their roots back to Loyalists who had left the United States out of persecution, to British emigrants hoping for a better life in Canada, or even to Americans who had taken advantage of land offers here. They had built a new place for themselves, far removed from any thoughts about the politics of nations. While the Loyalists never forgot about their persecution in the Thirteen Colonies, they recognized they had a lot in common with settlers on the American side of the river. The Thousand Islands had already become a sort of regional community where men and women worked and traded on both shores. Marriages across the border were not uncommon. There may have been a boundary line somewhere down the middle of the river, and there was a certain amount of national pride and rivalry, but hostility wasn't in the equation.

The Americans had won their independence from Britain some four decades earlier, yet they still suffered from trade embargoes and from losing their ships crews when they were captured and pressed into the British navy. The United States finally declared war on Britain in 1812, but was under no delusion that they could carry that war to England itself, or that they could win the war at sea. Canada, Britain's colony in North America, could be the only objective. The war was declared somewhat reluctantly: just over a third of the American House of Representatives voted in its favor. Those in upper New York State knew they had the most to lose, since much of their trade was carried on with Upper Canada. Eleven of their 14 members voted against war. The United States could hardly afford a war — the young country was already considerably in debt. And Britain was certainly not itching to go to war again in the Americas either. The British were in a very serious life-and-death struggle with Napoleon's armies in Europe, and could not spare any troops for such a conflict. As it was, though, war hawks in the United States won the vote to go to war.

The opening salvoes of the war were fired in the Atlantic in the interests of protecting American shipping. Shortly afterwards, the action spread inland, first to southwestern Ontario, in the Windsor-Detroit area. The conflict soon spread to the St. Lawrence River because this was the only route of supply available to the British. In the old days, the French had been able to bypass the insecurity of the river by taking the northern route up the Ottawa and French Rivers. In the 1800s, however, the enormous quantity of supplies necessary to the burgeoning settlements of Upper Canada left no alternative. Yet the Americans had various options open to them because of the number of rivers that could carry them north to Lake Ontario, Lake Erie, and the lower St. Lawrence.

Since neither side had really considered that there would be a war fought in this region, there were only a few relatively isolated points of defence along the border. These were mostly holdovers from the days of British-French conflicts, such as at Kingston's Fort Henry, Fort Wellington at Prescott, Fort Oswego on the south shore of Lake Ontario, and some minor fortifications at Fort Haldimand on Carleton Island and at Ogdensburg. Because the Thousand Islands region was essentially on a route and not a specific target, actions were largely limited to skirmishes and maneuvers. The numerous passages through the islands meant that the defence of communities and river shipping alike would be difficult. At the same time, the geography lent nicely to spying, smuggling, and covert traffic.

From the beginning, the British were in the best position, from a naval point of view, to command the waterways. The war had been declared on June 18, 1812, and within days, the British had ships on patrol on the river. On June 29th, a fleet of eight small boats tried to break free of the harbor at Ogdensburg and make their way to Lake Ontario. A man named Dunham Jones, who lived on the Canadian shore west of Maitland, saw the boats heading up the river. He realized what an advantage this could be for the

Americans, and quickly got a group of men together to give chase. The fleet was overtaken at the islands off Brockville. Two of the American boats gave up without a fight, and the others turned and fled back to Ogdensburg. The two boats captured were driven aground on one of the islands and burned.

A new British warship of ten guns, the *Prince Regent*, sailed down from Kingston to anchor off Prescott. She was joined by two more armed ships, and together they kept watch over the situation in that part of the river. An American three-gun schooner, the *Julia*, sailed down from Sackets Harbor and the first naval battle on the upper St. Lawrence began. At anchor and with all sails furled, the crews of the *Julia* and those on the two British vessels, the *Earl of Moira* and the *Duke of Gloucester* that had pursued the *Julia* from the lake, opened fire on each other. The cannonade lasted for an hour and a half, but neither side had much experience, and little damage was done to any of the boats. The following morning saw the *Julia* making her way into Ogdensburg, and the *Earl of Moira*, which had a few cannon balls go through its hull, put ashore at Brockville. There, the guns were taken out of her and added to the defence of the town. As it turned out, all of the American vessels were able to get to Sackets Harbor and were outfitted for the war before a short truce was arranged that summer.

Attack on Fort Ontario at Oswego, 1814.

Over the next two years, there were a number of what might best be called 'incidents' in the Thousand Islands. The majority of the big battles were fought elsewhere, although there were important naval engagements in the east end of Lake Ontario because two of the main installations, Fort Henry at Kingston and the American base at Sackets Harbor, were located there. Both had sizeable ship-yards, where most of the Great Lakes fleet of both sides were built. While Carleton Island had been a major British stronghold during the years of the War of Independence, it had become a very minor outpost, and actually lay in U.S. territory, by the War of 1812. As such, Fort Haldimand on Carleton certainly wouldn't have been considered a prime military target in the opening days of the war, but it was no doubt the source of indignation and an irritant to the local American citizenry. And so it was that shortly after war was declared, Abner Hubbard and two other Clayton-area men took it upon themselves, as the American Colonel Jacob Brown later wrote, to capture the old fort. A sergeant, three invalid soldiers, and two women were probably the first prisoners of war in the Thousand Islands. Fort Haldimand's name was changed to Fort Carleton thereafter but the fortifications themselves were never redeveloped.

For some reason, the Americans assumed that the village of Gananoque had stores of military supplies, including a vast hoard of munitions. These were to be the objective of Captain Benjamin Forsyth who made a daylight attack there on September 20, 1812. Forsyth and his 95 men marched to the town from the west, hoping to alarm the villagers by this approach. Alarmed they were indeed, but they got themselves organized enough to meet the Americans with a volley of musket fire. The shots all went wild, and Forsyth took this opportunity to charge the crowd, chasing them across a bridge, presumably over the Gananoque River. Just in case the Canadians recovered their wits and decided to return the attack, he burned the bridge.

A *British blockhouse on Chimney Island.*
(National Archives of Canada C11653)

There was supposed to be an understanding, an unwritten civility of those days, that private property would be respected during such raids. Some of the soldiers must have forgotten this when they barged into the home of Gananoque's leading citizen, Colonel Joel Stone. Perhaps it was just the excitement of the moment, but one of the soldiers fired into the ceiling when he heard sounds from the floor above. Mrs Stone, who had just hidden the family gold in a barrel of soap to avoid its theft, was shot in the hip and let out a howl of pain. The soldiers thought they had killed Colonel Stone himself and fled the house. Meanwhile, the rest of Forsyth's men succeeded in finding a small cache of arms, with about a thousand rounds of ammunition and some 40 muskets. These were taken back to their boats, but they reported to their commander that other provisions were left behind and burned. The townspeople held the opinion that the Americans overstated their success, and that the stores burned had little value at all, just a few blankets and an already spoiled side of beef.

Now stationed as commander at Ogdensburg, Captain Forsyth crossed the river by sleigh on the night of February 6, 1813, with 200 soldiers and volunteers. The foray was more of a retaliation than an attack because the British had gone to the American shore a number of times to arrest deserters. Forsyth's small force flanked Brockville, and the main body of troops went to the jail at Courthouse Square. They released all of the prisoners, except a convicted murderer, took several prominent residents as hostages, rounded up some 140 guns, two barrels of ammunition, and a quantity of stores before returning to Ogdensburg. A couple of hours later, two officers from Fort Wellington crossed to Ogdensburg and arranged for the parole of the Brockvillians.

Tempers flared over the raid on Brockville, and before long, the battle was carried to Ogdensburg. It seems that all that was authorized was a show of force, with British troops marching in front of the town on the ice, but, seeing no resistance, they continued on into the town. Only a couple of cannon shots were fired as the Americans withdrew to the woods. However, when asked to surrender, Forsyth said he would prefer to fight. The battle became more of a chase, with its conclusion near the village of Black Lake. Several of the Americans were killed and wounded, and 52 were taken prisoner and sent down river to Montreal.

The chimney, rebuilt since the War of 1812, and earthworks are all that remain of the Chimney Island defenses.

These raids back and forth on the river upset the relations enjoyed by most of the people on the two shores. Hospitality gave way to vigilance, and communities built defences along their waterfronts. Even at the flour mill at La Rue Creek, a little west of Mallorytown Landing, soldiers kept a watchful eye on the river from rifle pits near the shore. By the same token, there was an amount of resentment for the war's intrusion on the economy of the region. The mill owned by Billa La Rue, for example, was instructed to spend almost all of its time grinding for the troops at Kingston. La Rue was left to grind wheat and corn for the local folk at night and on Sundays.

There were some other types of trade that were driven underground and were preferred to be ignored by both sides. The Americans needed horses and the British enjoyed their beef. If there wasn't enough supply on one side of the river, then it had to come from the other. In fact, the commander of the British troops in Canada, Prevost, suggested that as much as two-thirds of his supply of beef came from the New England States and from New York. It would seem that contraband was more easily moved than troops. Horsethief Bay, west of Rockport, supposedly got its name as the place where horses were herded into the river to swim across to Hill Island and travel a well-worn path over to Wellesley Island. Watch Island, just off the north shore of Hill Island, was a place where patrols would look out for such clandestine activity, but it seemed like everyone was watching in the wrong direction. After all, this sort of 'free trade' or underground economy wasn't exactly new in the region. For years prior to the war, potash, for example, could be bought from Canada for a much lower price than was the market rate in New York State, but taxes would have to be paid if it were brought across the border and declared to customs agents. And so, on winter nights, after the river had frozen in, sleigh-loads of potash were drawn over the ice, often pulled by teams of white horses on runners that barely hissed on the snow, to landing points remote from any customs dock.

GUNBOAT PATROLS

If there could ever be anything picturesque about the war, it would have been the image of navy bateaux and gunboats rowing, drifting, and sailing through the islands. While the crews would have been ever-vigilant, the vast majority of their passages were peaceful. Supply convoys were very frequent on the river, from spring break-up to the December freeze. Troops on land saw little action throughout the war, but those on the river were constantly busy transporting, not just military equipment, but all things related to the supply and economy of Upper Canada. Roads in those days were more of a hardship to travel than a convenience: they were scarcely more than muddy and boulder-strewn tracks, with rows of logs, like corduroy, across the worst of swampy sections. The river was the true highway to the heart of the continent. In order to best protect the vital river corridor through the Thousand Islands, the

British formed a 'Gunboat Establishment', with nine gunboats that would work in groups of three to escort cargo-laden Durham boats and bateaux up and down the river. The passage on the river was broken up into stages with prearranged meeting places at Cornwall, Prescott, Brockville, Chimney Island, Gananoque, and Kingston. To protect town and countryside, blockhouses — squarish fortified buildings of stout timber — were built at Brockville, Chimney Island, and Gananoque.

The gunboats themselves were a very practical type of vessel for the Thousand Islands. Such boats had been used by navies in Europe for quite some time and the design was easily adapted to this region. They were generally low, wide, flat-decked and shallow-draft vessels and varied anywhere from 12 to 15 meters in length and three to five meters wide at the beam. Since wind and current seem to go in the right direction only half the time at best, the gunboats would have to be rowed or poled when they could not be sailed. They needed to draw as little water as possible for searches among the many channels, to be stout enough to haul cargo if necessary, and stable enough to carry one or two deck-mounted small cannon. They were built of local woods, as were all ships of the day, and planked with pine over heavy ribs of white oak, with hand-forged iron nails, often clenched over iron washers, driven through the planks.

A gunboat incident in the Thousand Islands happened not far west of the Chimney Island depot. A British gunboat was escorting a flotilla of 15 bateaux upriver. The convoy was on the way from Montreal to Kingston with 250 barrels of pork, 300 bags of bread, and other goods — apparently crooked suppliers in the Kingston area were blamed for short supply locally. The boats were intercepted by two American gunboats, each mounted with one cannon, just like their British counterparts. The supply caravan was captured, having been outnumbered by the 70 soldiers on the American gunboats. Somehow, the word quickly got to Kingston and three gunboats were sent downriver to recapture the convoy. They arrived at the entrance to Cranberry Creek, at Goose Bay on the American shore, too late in evening of that July day to press ahead with the attack. This gave the Americans time to set up an ambush.

The British gunboats, now four in number, were poled up the creek only to find their way suddenly blocked by trees felled across the water. The creek was marshy at this point, and so the soldiers could not easily wade to shore through the muck. The trap was sprung. The British were caught out in the open while the Americans had the cover of the surrounding forest. With a staggered 'crack' of muskets, an intense fury of balls whizzed through the air, thudding into the ships' hulls and whining in ricochet. A pungent pall of gunsmoke drifted from the trees over the marsh as the British shot back, firing the gunboats' cannon where they guessed the enemy to be. Amidst the uproar and confusion, the British managed to get their boats back down the creek to the safety of the open river. From the propaganda of the day, the account of the British was that seven of their soldiers were killed and 17 wounded, while the other side claimed that over 70 of the British were killed and at least that number again were wounded. In any event, the Americans decided it was best to quickly move on and make their way with the captured boats and supplies to Sackets Harbor. However, they ran into trouble themselves out in the lake when the British ship, *Earl of Moira*, rounded down on them. To avoid being taken, they sank most of the boats, and so the supplies were lost to both sides.

After that incident, the British stepped up their vigilance on the river and it was assumed that all would then be safe to pass. That wasn't a good move. An American by the name of Gregory was waiting for the opportunity to ambush something British. He had three small sailboats and a group of men hiding in a cove near the head of Tar Island, just a little east of the village of Rockport. Several larger fleets of boats passed by before a chance came along. The British gunboat *Black Snake* drifted downriver, her sails barely filled by a light breeze. Captain Landon and the 20 men aboard her didn't expect trouble, especially on this part of the river. When Landon saw the three little sailboats coming up the river towards him, he thought that they must be British gigs and so went out to meet them in a skiff lowered from the gunboat. The surprise was so complete that Landon was actually aboard one of Gregory's boats before he discovered what was up. Gregory, delighted with his prize, rowed off with the *Black Snake* in tow. They didn't get far, however, before the tables turned again. Another British gunboat saw what was happening and set out in pursuit. The Americans couldn't make fast enough progress and so scuttled the *Black Snake* near Deer Island and then were able to outpace the heavier and slower gunboat. The *Black Snake* was later raised and returned to service after a refit at Kingston.

The gunboats of the Thousand Islands again had an important role in a battle that was one of the pivot points of the War of 1812. In the fall of 1813, U.S. Major General James Wilkinson, a holdover and controversial character from the Revolutionary War, was sent to Sackets Harbor to take charge of the American forces along the St. Lawrence and Lake Ontario. By now it was abundantly clear that Montreal had to be taken. This was where the shipments originated that kept inland Canadian posts supplied. The plan was relatively simple. Wilkinson was to gather the troops from garrisons on the lake and sweep downriver to Montreal. Another force was to march north from along Lake Champlain, and together they would overrun the smaller British force. As it happened, Wilkinson and his counterpart and personal foe on Lake Champlain, General Hampton, were better procrastinators than decision makers. Fall gave way to the chill and rains of early winter. Wilkinson had assembled most of the troops from around the lake, leaving few to defend their posts. Finally the decision was made to pass on an attack on Kingston but to set out down the American side of the river. Much of the force of some seven thousand boarded a collection of whatever would float to make what should have been a quick dash to Montreal. Their progress was slowed by having to wait to keep pace with the remainder of the troops marching along the shore.

Isaac Chauncy, Commodore of the United States Navy, had tried to bottle up British ships to make for a safe passage of the troops on the river. Such a thing is easier said than done in the Thousand Islands. Several of the British gunboats followed Wilkinson's force, constantly ambushing them and firing on them with their cannons. Making its way on foot along the shore, the army was regularly harassed as well, and had to spend some time beating off a British raiding party at French Creek, where the village of Clayton stands today.

Wilkinson fretted about getting past the guns at Fort Wellington, still having to contend with the British gunboats and their gunners that pursued his force, when he learned that Hampton had been outmaneuvered before he got anywhere near Montreal. Meanwhile, British troops from Kingston were closing in from behind and the rumors spread that their numbers were overwhelming. Deciding to press on downriver, Wilkinson found himself boxed in by the rapids to the east and the British army and gunboats to the west. An encampment was made near Morrisburg on the farm of John Chrysler, and it was here that a battle took place that brought the American advance to a standstill. Despite their superiority in numbers, bloody fighting halted the invasion, and the American force crossed the river to take refuge in a very poor camp for the much of the winter.

It had taken over three weeks to cover the short distance from Sackets Harbor to Chrysler's farm. The cold and damp wintry weather, the tension of constant ambush in the Thousand Islands, and poor decisions from the command all contributed to Wilkinson's rout. All in all, it was plain to see Montreal would not be so easily taken. While there would be other scuffles here and there in the region, for the remainder of the war, and indeed for the rest of history, there would never again be a large invasion force to disturb the tranquility of the Thousand Islands.

THE BROWNS BAY WRECK

Keel, hull planks, and floor timbers of a War of 1812 era gunboat preserved at Mallorytown Landing.

For years and years, local kids swam around and 'cannon-balled' into the river from the timber sides of an old wreck that lay on the sandy bottom off Patterson Point, on the west side of Browns Bay Park. The hulk had probably lain in those shallows for over a century before it came to the attention of archaeologists. This proved to be an exciting find! It was raised, soaked in preservative baths, and housed in a display building at Mallorytown Landing in St. Lawrence Islands National Park.

While souvenir hunters and shifting river ice had pried some of the hull timbers free, and the decks were completely gone, enough of the hull remained to track down its origins. From the dimensions and construction details, it was soon obvious that this was a British gunboat from the time of the War of 1812. The original planking of the hull, for example, had been done with copper nails, as navy specifications would have demanded, but later repairs were done with iron nails since by 1820 builders knew that this type of fastener would last the life of the hull in fresh water. Although hundreds of ships of war of all sizes and types had been built by both sides for the war effort, very few examples survive, even as wrecks. A search of the records kept by the Royal Navy seemed to narrow the list of possibilities down to the H.M.S. *Radcliffe*. This was the last boat listed as built at the Kingston naval shipyard, completed March 31, 1817, just a short time before the Rush-Bagot Treaty of 1817 was signed. The terms of this treaty limited the number of armed ships that both Britain and the United States could operate on the Great Lakes. As it was no secret to either side that such a treaty was being negotiated, shipyards rushed to complete and mothball whatever vessels they could. Fortunately, this turned out to be a lasting peace, and the ships and boats were never needed for service again.

Smaller boats, like the *Radcliffe*, could be very serviceable workboats in these local waters. The navy records show that the *Radcliffe* was kept and maintained at the naval yard for a few years, with planks and timbers replaced as needed. Interestingly, while other vessels were eventually listed as surplus, sold, or destroyed, this particular boat simply disappeared from any list. If this is indeed the *Radcliffe*, it somehow wound up out on the river to continue its life as a workboat. A few modifications were made to the original structure to make it better suited to its new role. Since the new owner wouldn't have had a large crew to row the ship to windward, a centerboard trunk was cut into the hull, just beside the main keel. This would permit a centerboard to be lowered to keep the ship from side-slipping when sailed up wind, but be raised when in shallow water. Deep scars and gouges in the on the surface of planks inside the hull suggest that this boat saw rugged use, probably carrying cargoes of lumber, rock, barrels, and livestock as it was worked through the islands on the river.

THE TALE OF CHIMNEY ISLAND

There are two Chimney Islands that figure in the story of the Thousand Islands: one east of Prescott, Ontario, where there once stood an island fort; and a second, just east of Mallorytown Landing. It is the second Chimney Island that has such an intriguing tale.

Chimney Island rises from the river as a whaleback of granite, nearly smooth except for a few places where the glaciers tore away chunks large enough to allow a little soil to accumulate. The island is almost alone in this part of the river, accompanied by a handful of shoals that make their appearance as tiny versions of the main island in fall and winter when the water level drops enough to reveal them. The island is far enough offshore that in the old days, it may just have been possible for an arrow to carry as far as the island from the shore. The only deep channel on the Canadian side of this part of the river runs along the south side of Chimney Island. If a boat that drew any more than a fathom of water were to pass too far to the south of the island, it would go aground in the huge area of weedy and rocky shallows. Between the island and the shore is a bar of sand making the water quite shallow, especially in the fall of the year.

Knowing the setting of Chimney Island is important in understanding the island's role in history. It happens that the island is a natural lookout and control point at this section of the river. The view extends in every direction and yet any traffic on the river must pass close by. The setting was certainly important in the military history of the Thousand Islands, but it may have been important, too, for the island's first white settler, a mysterious Frenchman.

There is so little soil on Chimney Island it wasn't likely that this settler would have wanted to build there for farming. Timber cutting as an industry hadn't started as yet, and there was little trapping for the fur trade in this part of the country, so he couldn't have chosen the island as a base for that livelihood. When he moved onto the island in that winter of 1780, United Empire Loyalists were just beginning the settlement of the Thousand Islands with their grants of mainland farms. He could have chosen just about any of the larger and more well endowed islands on the river. Perhaps, then, the vantage point of the island was at the center of it all.

Chimney Island in winter mists.

Two Métis hunters had built a hut on Chimney Island the previous fall, but that was just a temporary shelter. The Frenchman began to put up a log cabin, pulling the logs across the ice from the mainland. It would seem that he meant to stay for quite some time because in the early spring, he left for Kingston to get the lime needed to make mortar for a substantial chimney — the island's first chimney. He left the island again in May and returned by bateau with not only supplies and furnishings for the house, but a bride as well. She was apparently a very pretty woman, of mixed white and Indian ancestry. The two lived a quiet existence on the island, having guests on occasion, but it seems that no one ever got to know the couple well enough to learn of their origins or how they were able to support themselves.

One day in the fall of 1780, Enoch Mallory and Joseph Bark were hunting along the river when they saw flames and smoke billowing from Chimney Island. They hurried to the island to see what was happening, and found the Frenchman at a little cove on the south shore. A new tomahawk was buried in his skull, and his body was in a partially burned canoe near the water's edge. There was no one else on the island.

The murder was never solved. Although the two Métis hunters who had occupied the island before the Frenchman's arrival were tracked down and presented to magistrate Thomas Sherwood at the village that would become Brockville, they were able to prove that they were away from the region at the time. No trace of the woman could ever be found.

Quite probably the second chimney on the island was constructed partially from the rubble of the first. It was during the War of 1812 that the British realized just how vulnerable shipping was on the St. Lawrence and how effective ambushes could be in the Thousand Islands. They had lost supply boats to the Americans on the river, and because they had no southern water routes into the Great Lakes, knew that every shipment was vital. Defensive blockhouses had been built for the protection of Brockville and Gananoque, but they knew that another was needed in the Islands at a strategic control point where shipping could have some cover. That place was Chimney Island.

The Chimney Island blockhouse was built in the closing months of the war. The blockhouse was a sturdy log affair, set on the south side of the island and protected by a rampart of stone and earth. A stone chimney was to provide heat and a place to cook for the men stationed there, but apparently its design was so poor that smoke filled the blockhouse any time the fires were lit.

Despite the less-than-satisfactory chimney, the blockhouse was a busy focus for the troops on the river in this area. There was a camp and parade grounds set up on the high ground on the mainland north of the fort. No doubt this was a place where the residents of the blockhouse were happy to go to escape the draughty, cold, and smoky quarters on the island, but it was also close to a footpath, the forerunner to the Old River Road that linked the budding villages along the river. Access to the island from the shore was made easier by adding earth fill to the sandy shallows between the island and the shore. This land bridge gave rise to a second name for the place, Bridge Island.

Chimney Island had a deep anchorage to the east and, in most winds, lea side of the island. Three gunboats that patrolled this section of the river routinely stopped here. Convoys of bateaux could relax for a spell in the protection of the blockhouse. It could be said that the fortification was effective in its role because shipping along the Canadian channels of this part of the river was never disrupted through the remainder of the war.

The blockhouse was either taken down or it completely decayed in the years after the war. The chimney, however, stood for almost a full hundred years, tumbling down in a storm in 1913. It had become a local landmark and was missed by area residents. Finally, William Gilbert, a cottager on Tar Island, took it upon himself to have the chimney rebuilt. The structure that one sees atop the island today is impressive, but not of the same design as that from the War of 1812. Local stories would have it that this chimney is no doubt the best built of the three — and does not smoke.

SHIPS OF WAR

Towards the end of the War of 1812, an arms race of a sort was going on at shipyards in Kingston and Sackets Harbor. The ships built for war on the Great Lakes had become lighter and faster than their seagoing counterparts and yet carried more armaments. Weather on the lakes, while it could be downright nasty, was not so vicious are were ocean storms. Hull construction could thus be lighter. Voyages were shorter, meaning fewer supplies of food and water were needed, thus making more room for guns and sailors to man the ships.

As one side or the other added a ship to its fleet, the vessel was a little larger than its predecessor. The British naval commander at Kingston, James Lucas Yeo, had been able to control the lakes with his navy at the beginning of the war, but in the spring of 1813 the tables were turned. The yard at Sackets Harbor was able to launch two large brigs, more powerful than anything the British had. Now bottled up in Kingston, Yeo decided to launch a truly mighty ship, something of a floating fortress, capable of breaking any blockade and delivering men and supplies to any port on Lake Ontario.

The keel of the *St. Lawrence* had been laid by the spring of 1814. Despite being blockaded on the lake, teams of oxen, ironically including some hired out of Vermont and New Hampshire, were successful in bringing cannon, cordage, sails, and other rigging overland from Montreal. Transport ships with more men and materials managed to get through as well. When the *St. Lawrence* slipped down the ways on September 10, 1814, there was absolutely no doubt about its superiority — she was very near 90 meters along the keel. Normally, ships were built of ribs spaced more than a meter apart, but in the *St. Lawrence* the uprights were a solid wall of wood. Every fourth rib was white oak, the strongest of the woods, but in between an assortment of lumber was used — walnut, pine, hickory, ash — anything the carpenters could find. The hull planks were also of white oak, 30 centimeters wide and 15 thick. Some were more than 15 meters long, requiring true feat of strength to bend these planks into place, yet they were steamed and fitted around the curved lines of the huge hull. The inside of the hull was lined with white pine planks, giving the ship a virtually cannon-ball-proof hull almost 60 centimeters thick. The *St. Lawrence* carried 102 guns, 34 of which were 32-pounders, the size of cannons usually put on the battlements of forts on land. The ship was handled by a crew of 640. When she sailed out of the harbor that October, no one questioned that the blockade of Kingston was over.

Despite her awesome strength, the *St. Lawrence* did not quite live up to all of her expectations. She was to have been able to carry a considerable cargo in order to supply as well as defend the ports around Lake Ontario. However, she was a little overweight and with a small load drew over six meters of water. Any sailor on the lake knows that such a deep draft won't allow a ship into very many harbors, let alone safely clear the shoals and bars on what was then a poorly charted lake. A passage down the shoal-strewn river in the Thousand Islands would have been out of the question. The biggest asset of the *St. Lawrence* was just what Yeo had anticipated: the shear power of intimidation. The *St. Lawrence* never fired a shot in battle. All she had to do was make an appearance. On top of that, keel blocks were already in place for building a sister ship.

Meanwhile, over at Sackets Harbor, Commodore Chauncy caught wind that trouble was brewing. Shipwrights were set to work to level the playing field. Chauncy had persuaded his superiors to let him proceed with even larger ships. The keel for the first of

The St. Lawrence *was the mightiest ship of the War of 1812-1814.*
(National Archives of Canada 982.19.223)

three ships was laid in the winter of 1814–1815. The *Orleans* was to have been completed by spring break-up, but work stopped when news of the end of the war finally reached the navy yard.

None of these old ships of the war survive today. The hull timbers of the *Orleans* stood on the ways for many years, eventually becoming built into the pier. The *St. Lawrence* served as a floating barracks for many years and was said to be warmer than the stone barracks ashore. Her last years were spent as a floating dock for cordwood, fuel for the steamers that had made ships of wood and sail obsolete.

The end of the War of 1812 came on Christmas Eve, 1814, with the signing of the Treaty of Ghent in Belgium. News didn't travel very fast in those days, especially in winter, and it was several months before word reached all of the towns and forts on both sides. There were those who would be slow to let any last resentment die, but the peace accord was to be one that lasted. Surveys were soon jointly undertaken to clarify the borders, and any last worries over the threat of war on the Great Lakes were put to rest with the signing of the Rush-Bagot Treaty in 1817, which strictly limited the number of armed vessels on the lakes.

PATRIOT'S WAR

The end of the War of 1812 came at a time when there was a growing need for new settlement lands. Emigrants from Britain and Europe came to the Great Lakes region in ever-increasing numbers over the next 20 years. The second generation of Loyalists on their quiet and sometimes lonely farms found themselves with new neighbors. Villages sprang up, local industries grew, and even some of the larger islands were settled, at least for the summers.

The growth of the Canadian settlement didn't come without growing pains. Roads, mills, education, law and order, mechanisms for export and import, among many other infrastructure needs, were not keeping up to the expanding population. In both Upper and Lower Canada, the people felt that there was little representation of their needs and views. They could elect their choice to the Lower Assembly but real power lay with the Upper Assembly, composed of the wealthy and well-established, appointed by the British government.

While Upper Canada had a large population of United Empire Loyalists, there were now many other British settlers who were a little less sympathetic to the motherland — Scottish, Irish, and English immigrants who were either forced off their lands or left Britain to come to Canada in the hope, and in some cases based on government promises, that they would find immediate prosperity in the colony. These people were willing to voice their protests about poor service and supply in the young country.

The state of unrest was an opportunity seized by William Lyon Mackenzie to harangue the Parliament in Toronto about its oppression of the people. When Mackenzie failed to achieve what he wanted politically, he took his fight to the streets, but lost there, too. After fleeing to Buffalo, New York, he soon recognized there was a certain amount of sympathy for his republican cause in America. As it happened, quite a few New Englanders who had fought the British for independence had moved to upper New York state and along the St. Lawrence. They were often inclined to believe Mackenzie when he expounded that Canadians were willing to take up arms against Britain. An organization known as Hunters Lodges attracted numbers of rebels. This secret society had its origins in Vermont, where similarly disgruntled Quebecers had immigrated to merge their idealism with sympathetic Americans. Some citizens of Lower Canada also wanted change in the style and substance of government. The Hunters Lodges movement was easily exported into upper New York State, adjacent to Ontario. Before long, organizers talked themselves into taking action.

The first hostilities of the so-called Patriot War happened in the Niagara Falls area, where Mackenzie's group had taken Navy Island in the upper Niagara River as their headquarters. A small force of Canadian volunteers captured an American ship, the *Caroline*, bound for the island with supplies for the rebels, set fire to it, and then cut it adrift to go over the falls. That incident of December 29, 1837, stoked tempers on both sides of the border. Rumors flew that the next moves were to be at Gananoque and Kingston.

That came to be in February 1838. Elizabeth Barnett, an American woman who taught school in Gananoque, had become engaged to a young man there. While visiting friends back in Lafargeville, she heard tell of a gathering at Clayton where plans had been laid to attack and capture Gananoque. Barnett crossed the frozen St. Lawrence — a rather heroic task, since river ice is anything but reliable — to warn the townspeople. The militia and eager townsmen turned out, armed and ready to brawl. A barricade of logs, destined to be rafted to Montreal in the spring, was rolled into place at the river's edge. Wagon loads of ladies were bundled off to safer points inland. Excitement built in proportion to the amount of whiskey poured. Reports came in that invaders were gathering on Grindstone Island and that as many as 2,000 men with field guns and supplies were now on the move to Hickory Island, off the north shore of Grindstone, west of Gananoque. Scouts were sent out to gather more information. They went on horseback, following the Wanderer's Channel westward. Even though it was biting cold, the ice was not thick enough at a spot near McDonald Island and a horse and rider went through. Nonetheless, the rest of the group continued up to Hickey Island, but found that the Americans had abandoned their plans. This time, cold weather and lack of organization and discipline had saved the people of two neighboring border towns from a conflict that no doubt both would have long regretted.

The Rock Island Light, where pirate Bill Johnson served time after the Patriot War.
(Parks Canada)

Three months and many speeches later, there was another incident that could very well have caused a real war. A Brockville steamer, the *Sir Robert Peel*, pulled alongside a wharf on Wellesley Island for refueling and to spend the night. Captain Armstrong had been warned that armed men had been seen near the wharf that day, but he chose to ignore the fact and let the steam off the boilers. The crew and passengers had retired for the night when 22 armed men, dressed like Indians, stormed aboard. Like pirates, the rebels looted the ship and chased the passengers, in various stages of dress and night-dress, ashore. Their leader, William Johnston, ordered the *Sir Robert Peel* cast off from the dock and the boilers fired up since he intended to take the ship for his navy in the Patriot cause.

But the pirates lacked ability when it came to steam engines, and the *Peel* drifted downstream, where the swift currents above Alexandria Bay grounded it firmly on a shoal. In anger, Johnston torched the ship, burning it to the waterline. The rebels escaped into the night, yelling, "Remember the *Caroline*," feeling they had revenge for the burning of their supply ship at Niagara.

William "Bill" Johnston had been born a British subject, but lost any feelings of loyalty to Britain during the War of 1812. He was a grocer in Kingston at the time, and frequently visited to the American side of the river to court his wife. He was accused of spying because of his visits and was forced to leave his business. His sympathies turned to the United States, which he assisted through the remainder of the war by guiding their ships and gunboats through the Islands. The Patriot War seemed to be a custom-made excuse for Johnston to find revenge. 'Pirate Bill Johnston', as he became known, wrote to the local newspapers after the burning of the *Sir Robert Peel*, and, in addition to claiming responsibility for the event, proclaimed that he acted as the admiral of the Patriot naval forces. He also stated that his stronghold was on Fort Wallace Island, in Canadian waters north of Grindstone Island. The reaction of governments on both sides of the border was to put a price on Johnston's head and on that of his men. Johnston was not to be captured, however, until later that year, following the bloodiest event of the Patriot War.

The Battle of the Windmill took place at Windmill Point, below the town of Prescott in mid-November 1838. Still believing that Canadians wanted to be free of Britain, a group of Americans planned an attack on the fort town of Prescott. In what started out as a well-planned assault, the steamer *United States* was rafted together with two sailing ships loaded with men and equipment, and journeyed through the islands to Morristown on the New York side. There, the sailing ships were set free to carry out the attack. The landing at a Prescott dock was bungled, however, and in the indecision that followed, one schooner ran aground on the muddy shoal off the Ogdensburg waterfront. Pirate Bill Johnston, the pilot on the ship, rowed ashore. Over the next couple of days, attempts were made to free this schooner while the second anchored off the stout stone windmill east of Prescott and put a small army ashore. Back at Prescott, a small armed British steamer, the *Experiment*, fired on the grounded schooner, on the *United States*, and on another small steamer that attempted to pull the schooner free. Local militia and regulars attacked the invaders at the windmill and a stone house they occupied, but the stone walls were too much for their muskets and light cannon. Eventually, a larger force with 18-pound guns came down river from Kingston and the rebels were cut off from supply and escape. They were eventually pounded into submission, but not before several persons on both sides were killed and wounded. When it was all over, the invaders were sent to trial. The leader was hanged, several were sent to a prison colony in Tasmania, but the youngest, some only boys, were set free.

The invaders had wrongly thought that their attack would be supported by regular forces in the United States and by sympathizers in Canada. In fact, some officials in Ogdensburg had crossed the river to the windmill before the final battle to try to persuade the invaders to leave for home. Bill Johnston was one of those who did see the writing on the wall, and he decided the time was right to make his escape from Ogdensburg and head back upriver. He rowed west towards the islands, pursued by American justice, and took refuge on an island off Brockville. After reflecting on the fact that his boat had been taken and that he faced the prospect of being marooned on the island without supplies for the winter, he gave himself up. Johnston was imprisoned for a year in an Albany jail, and after his release, he returned to the river at Clayton. Perhaps 'imprisoned' was not quite the right word, though. Pirate Bill Johnston had become something of a folk hero in those frontier days along the border. Apparently, his cell door was seldom closed and Johnston was seen to come and go from the jail, often accompanied by his daughter. He even had the freedom to attend a play that was written about his exploits. Bill Johnston lived out his days on the river, doing a little lighthouse keeping duty at the Rock Island Light as a part of his sentence.

The Patriot War was not a hostility declared by either Britain or the United States, but rather a localized uprising that got a little carried away. A group of sympathetic Americans had wrongly presumed upon the will of a few of their rebellious Canadian neighbors. One would think that if the same sort of events that unfolded in 1838 had happened along any international border today, the consequences would be severe. Despite all that transpired in those days, the governments remained remarkably cool-headed.

OREGON CRISIS

To anyone sailing from Lake Ontario to the St. Lawrence River past the Kingston waterfront, it must seem like this town was the scene of considerable conflict. The bastion at Fort Henry and the Martello tower fortifications at Cedar Island, Point Frederick, Confederation Basin, and Macdonald Park give the impression Kingston was a strategic location in the 19th century. And indeed it was. The backdoor to the French stronghold of Montreal for two centuries, Kingston controlled access to Lake Ontario during the 19th century and received the considerable attention of military strategists who upgraded the fortifications for the needs of the day.

Fort Frontenac, long since crumbled to ruins by the time the War of 1812 broke out, was seen as a poor site to protect the assets of the region. Point Henry, where a tall hill commanded the view of the east side of the Cataraqui River, was a much better defensive position. Fort Henry began as a blockhouse built on that point in 1813, a modest structure compared to the massive stone fort that dominates the hill today. Any sailing ship rounding into the sheltered bay at the Kingston waterfront would have to heave to in front of the point, in range of the guns.

The St. Lawrence River still remained a dangerous waterway, subject to easy blockade. To solve this problem, the Rideau Canal was constructed in the 1830s to link Kingston to Ottawa, via the Cataraqui and Rideau River watershed, effectively connecting the Great Lakes to the St. Lawrence at Montreal via the Ottawa River. Fort Henry became the front door to the canal system. Still, the fort alone was not deemed to be enough of a fortification to protect the area from a large force that might materialize off the lake. The incident that spurred the final upgrade of the defences at Forth Henry and Kingston was the Oregon Crisis in the 1840s.

Although the Oregon Crisis was a west coast border dispute, there was enough unease in the east that additions, long planned, were put in place at Kingston. Four Martello towers were built, an adaptation of Italian defences at Cape Martella in Corsica that had proven so effective in cannon bombardments as to frustrate the British

Kingston's defenses in the 1870s (top), with Martello towers at Cedar Island, Fort Henry, Point Frederick (Royal Military College), and Kingston harbor. Cedar Island (bottom) and the Cathcart Tower there (left). *(Parks Canada)*

Admiralty. Stonemasons went to work, building the four squat and barrel-shaped three-story structures out of local limestone. As the experience in Corsica had shown, the massive stone walls would deflect any cannon ball and only be pounded tighter into place in the process. The top of the towers supported big guns on rails so that the cannons could track the approach of ships, and the two lower stories housed a small garrison, offering often cold and smoky accommodation. Rounded, thick-walled stone fortified rooms radiated from the base of three of the towers to protect soldiers firing through narrow slots in the stone walls at any enemy who could ever get to shore. Each of the towers was slightly different from the other, reflecting the site on which it was placed. Three of the towers were in line: one on Cedar Island, at the east end of the line of defence; one on Point Frederick, the current site of Royal Military College; and one on a shoal in front of the present City Hall. The fourth, named the Murney Tower, guarded the west edge of the community from a vantage point overlooking the lake. An elaborate system of defence, these fortifications have, fortunately, never been battle-tested.

The American counterpart to Kingston's fortifications is at Sackets Harbor. Although nothing so complex was ever built there, some construction was carried out every time a conflict loomed. The military architecture and layout of the old barracks at the outskirts of town reflects the chronology of these conflicts, through to the two World Wars of the 20th century.

The possibility of Canada joining the United States, by choice or force, gradually faded from the forefront of political life along the Thousand Islands border, but the quest for more representative government in Canada did not. Eventually, the British got the message and Canada became a dominion in 1867.

THE TIMBER TRADE

Between these border conflicts and skirmishes, both countries began to develop businesses and industries that would thrive once the sniping, verbal and ballistic, came to an end. It was as if the final guns of the Patriot War were the starting guns of growth and prosperity in the region. The first glassworks in Upper Canada operated for a short time at Mallorytown, and a brick kiln in the same village turned out over a million bricks. One of the first iron foundries in Canada, just inland at Lyndhurst, provided materials for to local manufacturing industries in Brockville and Gananoque. These industries had markets not only in the Thousand Islands, but were also able to export farm and household ironware abroad. Another primary industry in the early to mid 19th century was timber cutting and lumber milling. While most of the old-growth forest had been cut years before to serve the British navy, local cutting for the building trade and for fuelwood to stoke steamboats was still an important part of the area's economy. The source of supply for the big timber had moved farther and farther inland, but the route to the coast was still down the St. Lawrence River. Before long, timber rafting was elevated from the status of opportunity to industry. The focal point of that industry was to be at Garden Island, across from the city of Kingston, at the west end of the Thousand Islands.

Delano Dexter Calvin, born in Vermont, moved to upper New York State in the mid 1830s and went into partnership with a local man, Hiram Cook. They leased 10 acres of land at the east end of Garden Island, making barrel staves and forwarding them down river. Ira Breck, Mrs Calvin's brother, replaced Cook in the expanding business by 1855. The men had seen the opportunities in forwarding not only staves, but timber itself down to the port of Quebec City. They knew that the prices for lumber varied, but the cost for the services of getting that timber to market could be a fixed and dependable income. When they mastered the process of assembling the logs into rafts, their business boomed, so to speak. They didn't just wait for logs to drift their way but went up the lakes after them.

Calvin and Breck also launched a shipbuilding industry at Garden Island, specializing in types for hauling timber. Some had big doors that would open in the bow to let hardwood logs be pulled into the hold and had booms to lift the lighter pine timbers to be carried on deck. These timber-haulers plied Lake Ontario to harbors where logs were gathered after being run to the shores of the lake from inland forests. When the Welland Canal opened commerce to the upper lakes, the firm built vessels to fit the dimensions of the locks. Eventually, the Garden Island ships could make their way to the head of the lakes. Other specialty ships were tugs built to tow the huge rafts downriver faster than the current of the quieter parts of the river would normally take them, and then return all the gear — the ropes, canvas, cooking equipment as well as the crews — back to Garden Island.

The big timbers were often over 25 meters long and, even when squared, over a meter across at the butt. They were heavy and cumbersome. Rafts traveled the Great Lakes at very slow speeds but were exposed to considerable peril from storms that could break them up or drive them ashore. Steam ships were the most effective means of carrying the timbers down the deep waters of the lakes. Getting the timber downriver was another matter. Much of the length the river from Prescott to Montreal was tumbled by rapids. A canal had been built to bypass the falls at Niagara, but to build a sufficiently large enough canal on the St. Lawrence was an even more monumental task, not to be undertaken until the St. Lawrence Seaway was built in the mid 20th century. The only practical way of taking massive shipments of timber down the St. Lawrence was to run the rapids with the timber assembled into rafts.

Rafts were made in sections so they could be assembled into larger rafts that would drift or be towed on quiet stretches of river, and then be taken apart again into more manageable units for the rapids. The squared timbers were dumped or pulled from the freighters at Garden Island and collected in a bay at the island's sheltered east end, where they would be put together as a raft. First, buoyant timbers were organized in frameworks called cribs. Rows of squared timbers were floated into these and held together by heavy pegs, wedges, and twists of ropy saplings called 'withes'. Two or even three courses of timbers might be put into the cribs, depending on the buoyancy of the species of the tree. Hardwoods such as oak and walnut, too heavy to float on their own, were pulled atop the rafts, along the edges or down the middle. Groups of four cribs were assembled into units called 'drams', forming rafts that might be 100 meters long. These in turn could be joined with other drams, creating floating islands as much as 300 meters long. Cooking and sleeping shanties were set up on the rafts, and

Top: *Timber rafts as these were assembled at Garden Island.* (Antique Boat Museum)

Bottom: *Rafting timbers down St. Lawrence River rapids.* (Parks Canada)

Timber rafts running rapids on the St. Lawrence River.
(National Archives of Canada)

guides for big sweeps to help steer were set along the front and back edges.

The rafts were more than just so much downriver bound lumber. They were also cargo vessels of a sort, sometimes laden with livestock, barrels of flour and pork, and bundles of staves for barrels. They even carried passengers on occasion. There was perhaps a no more tranquil way to view the Thousand Islands than from the timber rafts as they drifted along at the speed of the current. There was only the occasional hazard, such as when a sudden squall might drive the raft upon a shoal. A few mammoth oak logs can still be discovered on the river bottom where they fell from a raft wedged apart on the rocks. The real danger lay beyond the Islands in the rapids that would heave and torture the rafts as they hurtled down the sluices of green water, guided by the powerful sweeps of expert oarsmen who strained every muscle to follow the bellowed commands of the river pilots.

ISLAND FARMING

While never a booming industry like the timber trade, agriculture grew stronger and more diverse as the first farms passed from generation to generation. Farms first cleared by the early settlers on the mainland soon began to expand onto the islands. During the summer months, small islands were handy places to pasture a few cattle or hogs. The river provided a natural fence to keep the critters from straying. For the fall round-up, animals were herded aboard barges and scows for the trip back to the mainland and market.

Some of the larger islands, including Grenadier, Grindstone, Wolfe, and Howe, had substantial enough areas of suitable farmland that year-round farming communities grew up on them. The people who ran these isolated farms were certainly self-reliant individuals. Field and forest supplied the greatest part of their needs. Narrow lanes threaded around the islands, linking neighbors and sometimes leading to a hall or one-room school that was the center of the island's social life. On occasion — and it would have been quite an occasion — the family would board a skiff or work boat for

Abandoned farms and old machinery on Grenadier Island.

a shopping trip to a mainland town. This was a rare chance to explore the shelves of dry goods and hardware stores, to treat oneself and find fresh news of the world. Such trips were most often confined to the ice-free months; crossing the river in winter was a peril not taken lightly.

Dairy farming was the chief source of income, but only a few of the largest islands, like Grindstone, had their own milk and cheese plants. Horse-drawn wagons would trundle with daily precision to these island processors or to the main dock, from where the milk was rowed to mainland plants at Rockport and Gananoque. Considering the many moods of the river, that would have been more chore than pleasure for much of the year. Some of the milk plants had scows that made the journey to the islands. One such boat traveled the route from Rockport to Grenadier, using the township dock on the island's north shore. The abandoned hulk of this raft still lies off Grenadier's shore, just west of the old cement dock.

Over the years, the promise of these island farms has diminished as some farmers realized that their farmland did not have the capacity to make a good living. It takes large expanses of fertile lands to support strong farms. The rocky soils that had first discouraged settlers did not have the productivity needed for the demands of growth. Some smaller farms and areas of marginal farmland were abandoned. The small cheese plants and dairies faded as well. The plants at Grindstone Island, Gananoque, and Rockport, long fixtures in the Thousand Islands, are now gone. A faster paced industry, tourism, came along to fill the gap. Some of the island farmhouses were converted into summer homes. Farming as a way of life in the Thousand Islands is fading from memory, with fields once cleared by ambitious settlers disappearing into the new growth of woodlands. Fortunately for all who enjoy the natural scenery and rustic character of the Thousand Islands, this region has remained true to its rural roots. Much of the woodlands and mainland farms remain untouched by the industrial growth that swept through the heartland farther west on the Great Lakes.

MARKET HUNTING

There is another part of the story of the Thousand Islands that is told from Ice Island. It's a story from the 'market-hunting' days of the late 1800s, when many bountiful regions of North America were seen as a source of game for big city markets.

"There is absolutely no truth," emphasized the old decoy carver, tapping a weathered finger on the table where some of his works, old and new, were displayed at the local rec hall, "that my grandfather shot that game warden. He probably wasn't even on the river that night." Stories about those long-gone days of market-hunting, to keep game on the menus of distant big city restaurants, are more rare to stumble across than tales from the Prohibition. "Fact is," he muttered, "it was just a warning shot meant to go over his head."

Ice Island, and many of its neighboring islands and shoals, lie in a large area of fairly shallow, plant-rich waters. By the time much of the spring break up occurs, enormous flocks of ducks and geese settle in this stretch of river for a period of rest during their long migration to northern wetlands and breeding grounds. This section of the St. Lawrence has been such a staging area for untold thousands of years, a perfect place for hunters to surprise the migrating flocks and make fair money in bringing waterfowl to the city markets. There are still old stone shelters called hides or blinds on Ice Island, and on some of the other small islands nearby, visible even from shore. Structures like these were built on many of the shoals in this and other parts of the river. Hunters would also drift down on the restless flocks in the cover of early morning darkness using low freeboard punts. Oversize shotguns, some practically small cannons, would thunder handfuls of lead shot into clouds of waterfowl as they took flight. Even when enlightened game laws were put in place to outlaw the spring hunts, in attempts to preserve rapidly dwindling populations, the temptation for some of the die-hard hunters was a little too great to resist. Game wardens had their work cut out for them in those days.

Soon enough game was no longer sufficiently abundant in the Thousands Islands to support market demand and hunting became a recreational sport, a component of the tourism industry that transformed the region in the second half of the 19th century.

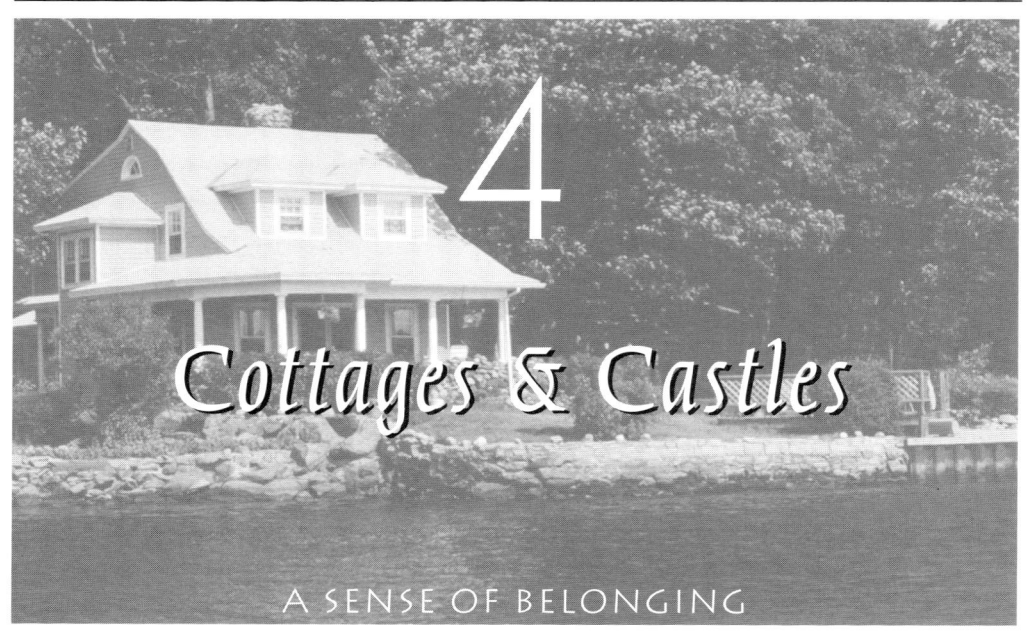

Cottages & Castles

A SENSE OF BELONGING

Cottage on the south shore of
Tremont Park Island.

She would never think of getting up that early on weekends back in the city. But here they were, on
'summer time' at the cottage in the Islands, just a few days out of the city groove, and morning came
as a pleasure. There was no alarm clock. There was no sound of traffic gearing up for the daily
momentum of business world life. The only hustle and bustle came from the branches of the pines outside the
screened and opened windows, where the chickadees blustered their way through the needles looking for tiny
beetles or little caterpillars that were inching their way to an early breakfast of their own. Perhaps the wake-
up call had come from the rustling of oak leaves as a gray squirrel leapt from one branch to another, or the tapping of the woodpecker on the trunk of the hemlock down by the boathouse. Whatever it was, the first silver
light of dawn was just enough to make it down to the kitchen and the morning ritual of coffee.

Next stop was the veranda. She pushed the wide screen door open and eased out onto the weathered floor,
setting the door quietly against the latch. On slippered feet, she shuffled over to a favorite the broad-armed
wooden chair and, cradling the warm mug, settled back to watch the morning unfold. There were no dramatic
moments, just the little things that could never be noticed in 'everyday' life: the heron gliding over to the dead
branch of the white cedar by the shore, a perch which allowed both a sun-warmed vantage point and a lookout

for breakfast that might swim by. The slow, regularly spaced roll of wake across the dusty purple-gray surface of the morning river, originating from a earlier-rising fisherman's boat. The orange and black flash of an oriole in the tops of the ash trees and the completely original, never-repeated melody of a catbird in the chokecherry thicket off to the side of the cottage. No two mornings were ever the same.

She had come to realize that her grandmother had enjoyed the same pleasures of the morning from the veranda's chairs and settees. As long as she could remember, the chairs hadn't moved more that the few inches necessary to sweep around the legs. Their placement gave the best views of the comings and goings on the river and of the setting sun. Like so many other things at the cottage, there was a sense of belonging — they had become a part of the culture and fabric of the place. Some traditions went even further than the furnishings. There were expressions of love for the place throughout the building. Walls had become galleries for watercolors and photographs that could only mean something to the people who shared in the experience of the place. Bits of beaver-chewed wood, stacks of colorful stones and pebbles, and iridescent bird feathers adorned mantle and shelf. Eclectic selections, or more correctly, collections, of magazines and novels stood ready on shelves for fourth or even fifth discovery on a rainy day. Closets were full of favorite hats and coats and shirts and shoes that would never have survived a spring cleaning frenzy back home. Kitchen drawers and cupboards are the refuge of cutlery, dishes, and pots and pans that had been demoted from the kitchen at home, but were too good or too useful to be just given away. Such is the way of summer places. They are comfortable in their treasured informality.

People throughout the world are drawn to places where they can find sanctuary for the soul, refuge from the pace of the working world, or simply a place where they feel comfortable with their own character. Whether grand or rustic, the camp, the cottage, or summer home is where the heart belongs. The Thousand Islands is just such a place.

THE 'GOLDEN AGE'

"He would have drowned sure as the world if he hadn't been pulled out," Captain Elisha Visger commented as he pointed out to a reporter the spot along the shore where President Grant had tumbled into the river during his visit to the Thousand Islands back in the spring of 1872. The correspondent from the *Watertown Daily Times* was conducting a casual interview on board a regular tour on the *Island Wanderer* with Visger. They had now come most of the way around the route from Clayton, north to Gananoque, down the channels past Hill Island and back up river past Alexandria Bay, and were now passing along Pullman Island. It was this island that had sparked the memory of the President slipping unceremoniously into the river. There were no doubt a few agents who got a 'soaker' in the process of the rescue.

Ulysses S. Grant was campaigning for his second term as president when he was invited to visit the Islands by George Pullman. This was a lot more than a casual invitation: it was part of a promotional plan to raise the profile of the Thousand Islands among the leagues of city gentlemen who might be enticed to develop properties in the region. As it happened, Pullman was a friend of Andrew Cornwall and John Walton. These two men were executors of the will of the late Azariah Walton, who had come to own almost all of the American islands by the year 1850. The elder Walton's original interest was the timber on the islands, but he soon became aware of the growing number of sportsmen who were drawn to the area by the superb fishing. In fact, in 1854 a naturalist from New York by the name of Seth Green asked Walton if he could purchase an island and build a cottage on it in order to study the fish of the region better. Although Walton thought the request was a bit unusual, he consented, selling Green the island now called Manhattan, just north-east of the village of Alexandria Bay, for the sum of $40. Sale of the islands was slow, however, and by 1856 only seven deeds were transferred. Even so, Cornwall and Walton speculated on how much money could be made if one day the Thousand Islands were to be 'discovered'.

Recognizing how important it was that homes be seen on the islands, Cornwall and Walton specified in their sales agreements that buildings must be erected within two or three years. They were also astute enough to realize that no one would be attracted by completely lumbered islands, and decided to keep every second island in its natural state. This in itself was an important turning point for the fate of the natural environment of the Thousand Islands.

George Pullman, whose enterprise had developed the railway sleeping car, purchased Sweet Island from Cornwall and Walton. He saw the potential for his business if the islands were further developed because his cars would be needed on the trains carrying 'tourists' on the long trip up from the cities. "What we need to do, Andrew," he is reported to have said, "is to make much of the General's visit here, and it will advertise the Islands as no other thing we can do. To have the President of the United States as our guest is quite an honor." And he was quite right.

The event of Grant's visit was carefully orchestrated. Because the trip coincided with an annual convention of the New York State Editors and Publishers' Association that happened to be held in this region, there was wonderful coverage of the President's visit. A train took 200 visitors to Cape Vincent, where they boarded steamers for excursions on the river to Clayton and Pullman's island. Receptions, bands, flags, and flowers provided a lavish backdrop against which the reporters could not write but the most glowing praises of the region. Their reports heralded the dawn of a Golden Age in the Thousand Islands. It was not long before many others came to build their summer retreats.

By the late 19th century it was clear that the Thousand Islands would become one of the great resort areas of the world. Not only was the landscape beautiful, the region was accessible by rail. The Thousand Islands were adopted by families who in some cases have a history of five or more generations at the same summer place.

As the popularity of the Thousand Islands grew, so too did the number, size, and elegance of the hotels. Hotels could be found in all of the communities on both sides of the river and on several of the islands. Some of these resorts were enormous, even by today's standards, with their own power generation, greenhouses, boat liveries, bakeries, and fire stations. The Frontenac Hotel on Round Island off Clayton, the Pullman Hotel on Grenell Island, the Columbian Hotel at Thousand Island Park on Wellesley Island, and the Thousand Island House at Alexandria Bay made elegant and

Camping in the Thousand Islands in the 19th century. (Parks Canada)

lavish statements about the new-found prestige of the region.

The Thousand Islands was not solely a playground for the rich. Far from it. The region attracted people from all walks of life. Even in the days when Walton and Cornwall were trying to attract buyers for islands, there were plenty of small inns, boarding houses, and hotels catering to sportsmen and their families who came from American cities to enjoy the natural riches and tranquil beauty of the area. A number of smaller hotels were built to keep up with the demand for accommodation by fishermen and tourists who wanted less pretentious and less expensive surroundings. The Ivy Lea Inn, built by the Visgers, the Thousand Islands Inn at Clayton, and the Grand View House on Wellesley Island are still standing, though the Tremont Park Hotel on Tremont Park Island off Gananoque is gone. The majority of the hotels, big and small, were mostly wooden structures so that they could be built as quickly as possible to take advantage of the rapid growth in tourism. Alas, the structures were vulnerable to time and, more often than not, to fire, which tore through the dry old buildings before the inadequate means of fire-fighting could save them.

In addition to luxurious hotels and more modest inns, all sizes and manner of summer homes were built by the burgeoning population of summer people. They wanted more than the casual remembrance of a few nights stay in one of the hotels. Those who had the means sought out land agents to stake their personal claim on an island treasure in the St. Lawrence. The Golden Age of the Thousand Islands will be remembered best for those magnificent island homes that were among the finest ever to be built in any resort area in the world.

President Ulysses S. Grant at Camp Charming on George Pullman's island.
(*Les Corbin Collection*)

115

Crossmon House and Thousand Island House at Alexandria Bay (opposite).

Skiff livery at Crossmon House (left).

Hopewell Hall, Castle Rest, and Nobby Island near Alexandria Bay (below).

(The Thousand Islands, James Bayne Company)

BOLDT CASTLE

Without any doubt, the summer place most identified with the opulent era of this region is Boldt Castle on Heart Island. Featured in practically every promotional brochure about the Thousand Islands, this castle is the showpiece of almost every boat tour of the region.

Boldt Castle in the early 1900s.
(National Library of Canada C27931)

The story of Boldt Castle has two beginnings: one with the island, one with the man. For its part, the island not only hosted a portion of the rich ecology of the region but features prominently in the most famous chapter of the Thousand Island's cultural history. In July 1871, E. Kirke Hart of Albion, New York, purchased what was then called Hemlock Island from the Waltons. Hart, who had a successful career in politics at the state and congressional levels, was also successful in business. He established the *Post Express* newspaper in Rochester and was president of a bank. Hemlock Island was considered one of the prime locations of the day as the island faced the town of Alexandria Bay directly across the channel. By the following summer, Hart had given the island his family name and had begun construction of a summer home. The $10,000 cottage, a considerable expense then, was built of wood on a foundation of granite. Tall and graceful, the home boasted a tower at one end. His ownership of Hart Island lasted about 22 years until his death in 1893. The family continued to own the island for another two years.

George C. Boldt's first visit to the Thousand Islands was in the same year as Hart's death, a weekend visit to join his wife, Louise, and their two children at the Thousand Islands House in Alexandria Bay. Having thoroughly enjoyed their first experience in the region, the Boldt's returned the following summer and, along with another family, spent the month of July on a chartered steam yacht, the *Sophia*. From the vantage point of the 23 meter (75 foot) yacht owned by Captain Harmonius W. Visger, George and Louise Boldt developed a deep attachment to the Thousand Islands and were especially attracted to Hart Island. The Boldt's purchased the island from Mrs Hart through a third party in the summer of 1895 for the amount it had taken to build the cottage.

George Boldt was not the type of man to be content with merely having an island from which he could enjoy the tranquility of the region. He was an extremely ambitious business man who, after emigrating from Prussia at an early age, came to the United States to make his fortune. After an inglorious start in a hotel kitchen and as a ranch hand in Texas, he went to New York City where he began his career in hotel management. Boldt quickly showed himself to be an excellent manager. In a move to Philadelphia, he was offered employment at the Philadelphia Club, and before long, married the club steward's daughter, Louise. Then began Boldt's meteoric rise in the hotel business. With the backing of Philadelphia Club members, he bought and converted a large home into the Bellevue Hotel in 1881. Seven years later he purchased a neighboring building to establish the Stratford Hotel and in addition managed other hotel operations in Philadelphia and New Jersey.

These successes caught the attention of William Waldorf Astor, who engaged Boldt as the manager of two hotels they planned together, New York's Waldorf and Astoria. George and Louise Boldt brought so many innovations in hospitality and service to these two hotels that their management style became legendary. Boldt's ability in business didn't end with hoteliery, but extended into the realm of investment as well. Throughout his life, he continued to amass both reputation and fortune. This was what enabled him to build as big as he could dream in the Thousand Islands.

Soon after buying Hart Island, Boldt decided to reshape the perimeter of the island in the outline of a heart, and so it was renamed 'Heart' Island. The original cottage was continually remodeled and expanded to the point where it became one of the most impressive 'cottages' on the river. Every year brought another building project. A castle-like granite tower for poultry, boathouses, apartments and a clubhouse for servants, and Alster Tower, reminiscent of a tower Boldt had seen in Germany, were constructed on the extensively landscaped island. An archway that resembles the Arc du

The Dovecote at Boldt Castle, for poultry and pigeons.

Triomph in Paris framed the entrance to a lagoon at the west end of the island, through which guests arriving by boat would pass. Six guest houses were linked by stone buttresses. A 'Roman' swimming pool at the front of the main house and a power house at the east end of the island, complete with clock and bell towers, all added to what had become the most elaborate estate in the Thousand Islands. Even so, George and Louise Boldt had a more extravagant plan in mind.

Perhaps it takes once-in-a-lifetime events like the turning of a century to spawn incredibly grandiose projects. In any case, in the summer of 1900, work began to remove the 80-odd room Hart cottage off Heart Island. In its place, George Boldt intended to build a massive granite castle as a monument to the love of his life, Louise. The six-story, 127-room Italian Renaissance-style structure would house everything one could ever need to make life in the Islands most enjoyable. There would be guest rooms worthy of dignitaries, ballrooms and studies, and all of the facilities the many servants would need to serve the family and friends. The Boldts took a hands-on approach to their project. They worked closely with the architects, owned the quarry on Oak Island where the granite was hewn, had sand and gravel drawn from their own pits, and directly employed the army of quarrymen, stonecutters, carpenters, electricians, plumbers, masons, landscapers, boatmen, gardeners, bakers, and other tradesmen who would make their dream a reality.

By the winter of 1902, the castle rose above the island's trees to grace the vista from Alexandria Bay. The building was originally conceived to be built from rough and moss-covered stone to give it the appearance of age. The stonecutters' work was so very good that the form taken was much more finished than would appear in nature. The work was proceeding rapidly, and marble fixtures and wood from around the world were being delivered, ready to be put in place when the time came. Then, on January 7, 1904, Louise Augusta Kehrer Boldt died. Although only 42 years old, she had been in poor health for some time and apparently died of heart failure. The reason for building the Heart Island castle suddenly ceased to exist. George Boldt telegraphed to Alexandria Bay and ordered that all construction be stopped. The workmen dropped their tools and the dream of two lifetimes was never taken up again.

While the story goes that George Boldt never set foot on Heart Island again, he did not lose interest in the Thousand Islands. In fact, his projects in the area were far from over.

Cottages at Ivy Lea Village.

The water tower at Calumet Island *(top left)* is visible for miles.

Dark Island *(top right)*, on the channel east of Alexandria Bay.

Jordstadt Castle on Dark Island *(bottom left)* was Fredrick Bournes' 'hunting lodge'.

The restored powerhouse at Boldt Castle *(bottom right)*.

Cornwall House at Rockport (top).

Victorian cottages at Thousand Islands Park (bottom) seem to compete for the most and best gingerbread trim.

The roof of Napoleon's Hat on Hay Island (top) was built on the ground and lifted into place.

Niagara Island summer home (bottom) duplicates a design for a Mediterranean villa.

Stately "Seven Gables" (opposite, top left) on Pine Island.

Zavikon Island's bridge (opposite, top right) is one of dozens in the Islands.

Rural traditions set the pace of life in the Thousand Islands (opposite, bottom left).

Surveyor Island's quaint cottage (opposite, bottom right) built in the late 1800s.

Evenings' golden light accents a gazebo (above) in the Sport Island group.

Lighting the Way: Three Sisters Lighthouse (top left), south of Grenadier Island.

Cross-over Light marks a critical turn in the Seaway (bottom left) between Alexandria Bay and Brockville.

A spar in the strong currents at Ivy Lea (top right).

Channels markers like this one above Ash Island (bottom right) are sometimes used by ospreys as nest platforms.

The range mark and light at Hillcrest.

Breaking the morning stillness at the
Thousand Islands International Bridge.

The Boldts had owned considerable property on Wellesley Island, where Boldt had spent another small fortune on farms and gardens to supply his kitchens, stables and fields for polo ponies, tennis courts, guest buildings, docks, canals and buildings for yachting, and other such things as might be needed to make life in the Islands more suitable. There was also a huge boathouse built on Fern Island, just offshore from Wellesley, to house and service his 60 or so watercraft ranging from skiffs and workboats to luxury yachts and speedboats, including the houseboat, *La Duchesse.*

George C. Boldt died on December 5, 1916, leaving his considerable estate to be divided evenly among his two children, George Jr. and Louise Clover Boldt. Even before Boldt's death, others began to dream of what the castle might become. Various thoughts and schemes included a 'White House' north and a luxury hotel. None of these came to be, but all the while the island and castle had become one of the favorite tourist attractions in the Thousand Islands. Ownership of Heart Island and the Wellesley Island properties went through various hands in the 1900s, but today, the Boldt Castle property is owned by the Thousand Island Bridge Authority, a non-profit organization which has undertaken the repairs of the buildings and developed the interpretation programs offered during the summer months.

There are no plans to complete Boldt Castle, but every summer there is work done to repair damage caused by vandals and weather over the years. Some of the rooms have been restored to the stage that shows how the plaster work, marble, and wood would have been used during construction. Crates of fixtures, mouldings, and even dishes were on the property at that time, though much of this was carried off by looters over the early years. After all, the castle was nearing completion when the order was given to stop.

THOUSAND ISLAND DRESSING

A restored parlor at Boldt Castle.

While Thousand Islands dressing may be one of North America's favorite salad toppings, to this day no one can verify its origins. For years and years the myth has been perpetuated that George Boldt's most famous chef at the Waldorf Hotel, Oscar

Jorstadt Castle, north side.

Tschirky, whipped up this little gem while Boldt was entertaining guests with a luncheon aboard one of his yachts. Some years had passed before anyone actually asked Oscar whether this was the case. Surprisingly, the great chef admitted that he had never been to the Thousand Islands, nor had he concocted the recipe. Over the years, several persons, from innkeepers to grandmothers, have laid claim to the fame, but let's just say that no one is collecting any royalties.

Since there is no 'authentic' version of the dressing, one can imagine this condiment to be as fanciful and flavorful as one likes. However, here's a simple blend: mix a little chunky green relish (the green represents the islands) with a splash of red ketchup (for the flavor of these world-class sunsets), then add a dollop of mayonnaise (for the whitecaps on the river).

JORSTADT CASTLE

Another impressive building in the Thousand Islands is Jorstadt Castle, sometimes called Singer Castle. The massive rough hewn-granite walls and red tile roofs of this 'castle' tower above the bluffs of Dark Island, in the middle of the river along the main shipping channel between Mallorytown Landing and Chippewa Bay. While not quite as large as the castle on Heart Island, Jorstadt is imposing and has been lived in during the summers through most of its history.

Frederick G. Bourne, a vice-president of the Singer Sewing Machine Company and commodore of the New York Yacht Club, purchased Dark Island from Andrew Cornwall in 1905. The name 'Dark Island' was more descriptive than sinister in intent. An early name for the cliff-sided island was Bluff Island, but it became more commonly known as Dark Island because its big stands of dusky-green pine trees made it look quite dark from a distance on the river. Bourne had first visited the region as a sportsman. He built the castle as a surprise for his family, telling them that he'd had a summer 'hunting lodge' thrown up in the Thousand Islands for them to stay at during the summers.

In fact, Bourne had spent more than a half-million dollars in construction on the island, certainly a huge amount of money in those days. The main house is reminiscent of a European castle, with an entrance way of stone arches, steps, and fireplaces. Suits of armor, torches on wall brackets, and slate floors add to the castle atmosphere, as do the towers and turrets of the building. The many rooms and halls in the middle and upper floors of the building, however, are much more like one would expect from architecture of the early 1900s. Bournes 'hunting lodge' is comfortable and certainly well-appointed.

For many years, stories spread about 'secret' tunnels on Dark Island and passages in the castle. There in fact a number of passages in the main building, though not secret, that allow service to some rooms without disturbing guests in other rooms along the halls. Connected to a central stairway, these passage ways open into the backs of closets, an image that has sparked the imagination of storytellers. Other buildings on Dark Island include quarters for house staff, an ice house, a pump house, kitchens, a bath house, and, quite unusual for its day, an enclosed squash court.

Bourne, like several other wealthy summer residents of the region during this golden age, was a boating enthusiast who expressed his passion by having his own fleet that included race boats. He constructed two large boathouses on Dark Island and another on nearby Grape Island. The boathouse on the south side of Dark Island, built of stone and boasting large windows overlooking the river, was substantial enough to shelter several boats at a time. A large concrete pier protected the main boathouse from waves and wake while offering island guests a place to disembark. Bourne moored his various steam and gasoline powered yachts inside the pier, including the Artemis, the Dark Island, and the Sioux. The boathouse one on the north-east side of the island housed work boats.

Whether encountered upbound or downbound by boat on the St. Lawrence River, Jorstadt Castle on Dark Island is an unmistakable landmark. In the evening, the massive windows of the huge hall on the castle's east side reflect the orange glow of the setting sun with a brilliance that can't be missed by anyone driving along the Thousand Islands Parkway.

A massive lantern by the main door of Jorstadt Castle.

Jorstadt Castle's entry hall.

A DARK ISLAND TALE

The farmer tugged his heavy woollen coat closer around his neck and turned a shoulder to the cold, damp wind. It was his last trip over the frozen river for the day. One more heavy stoneboat to unload and he would turn the horses to the shore, and home. The horse's hooves were nearly silenced by the wind-packed snow on the ice but the iron-shod runners of the oak sleigh hissed and groaned across the snow patches between pressure cracks. It was a bitter cold spell in this winter of 1906 , but had made perfect ice for the job of hauling rock off old man Bourne's island. "Who would have thought," he mused "that someone would pay a man to build another shoal in the St. Lawrence River." The river was already strewn with rocks, and for every island that made it above the surface, it seemed that there were two more that did not.

And now here was the spot. Rock heaved and shoved from the stoneboat added to the growing pile of rubble. It seemed like muffled thumps and cannon-shot cracks of the ice grew louder as he added new weight to the heap, but the sounds came from all around, as much as from underfoot. The horses shifted from foot to foot, anxious to get along. He knew that this shallow place, away from the main channels, had thick and safe ice, but even so was a little relieved to turn the rig towards Mallorytown Landing. When the ice rotted in the early March sun, no one would be around to see or hear the rock plunge to the river bed. Perhaps he would return by skiff on some glass-calm day to see his new shoal just downriver from the Dark Island building site and to find as well if by chance a few bass or even a muskie had laid claim to these new rocky shallows.

CHURCH CAMPS

Campgrounds had been used as religious meeting places in the United States since the early 1800s. By the middle of the century, Baptist and Methodist church groups were making use of extended camp meetings to provide places where people could restore their physical and spiritual health, away from temptation.

The natural beauty and tranquility of the Thousand Islands were surely good for the soul. In the early 1870s, a number of these camps sprang up in the Thousand Islands at Summerland Island near Alexandria Bay, near Morristown on the New York State shore, at Butternut Bay 10 kilometers west of Brockville, and on the southwest end of Wellesley Island. One objective of the camps was to encourage social interaction between Canadians and American. The St. Lawrence Central Camp Ground at Butternut Bay and the Thousand Island Park on Wellesley Island were the largest and best organized of the Methodist camps. In their earliest years, most of the campers pitched their canvas tents on the bare ground, but it did not take long before amenities were added, including docks, dining halls, meeting halls, tents on raised platforms, and, eventually, boarding houses. While visitors could arrange for tents and rent everything needed for the outdoor experience, many people were inclined to think to the future and purchase the rights to particular campsites. Cottages were built and hotels opened. Before long, the campgrounds had the air of permanency, although no one owned the land upon which their tents were pitched or cottages erected.

Many of the old cottages can still be seen at the campgrounds, particularly at Butternut Bay and at Thousand Islands Park, far and away the largest of the camps, with many delightful summer homes. There is so much gingerbread, color, and trim detail on these Victorian-style buildings that they seem to be in contest with each other. Likely, when they were built, that was exactly the case, with local carpenters interpreting this style in ways that would best suit the climate and the view. These builders were often hired to construct every structure on the property for their summer clients, from dock to icehouse, from house to outhouse. Roofs were steep, eve overhangs were wide. Veranda pillars and railings were solid, windows were shuttered against winter storms, and beams, foundations, and fireplaces were much more massive than would be seen in city versions. The architecture in the church camps — indeed, in the Thousand Islands in general — is a solid, weather-sturdy form that has given generations of families not only pleasure in their appearance but also confidence in the home's ability to stand up to the seasons.

Now in private hands, the old church campgrounds and cottages are wonderful places to visit. It's easy to let your mind slip back through time and imagine what summer life would have been like here over a century ago because that special summer way of life is still enjoyed today.

Boarding the steamers at Thousand Island Park.
(The Thousand Islands, James Bayne Company)

LA VIGNETTE

In the Wanderer's Channel, just offshore from the St. Lawrence Islands National Park's McDonald Island, there is a small but extremely picturesque cottage called La Vignette. Just before 1900, a Boston architect named Frank Lent designed the cottage which gracefully steps across two small islands, known as the Sisters. The raised walking bridge forms an arch that links the cottage island with the boathouse island and frames a view of a cabin in the background. The same architect also designed the clock tower on Stone Street in Gananoque.

La Vignette cottage in Wanderer's Channel.

NAPOLEON'S HAT

Napoleon's Hat is the name given to the unusual cottage on Hay Island, south of Gananoque. At a glance, it's not hard to see where the idea for the name comes from. The cottage was built in 1913 for Gananoque businessmen. The roof was constructed like a boat hull, with roof boards bent and nailed to the curves of the specially sawn rafters. The structure was built on the ground and then raised up to have the walls put in place beneath.

BOSTWICK ISLAND

Bostwick Island, in the Admiralty Group at the west end of the Wanderer's Channel, was originally given the name Yorke Island in Owen's 1815-1817 survey of the Islands to commemorate Sir Joseph Sydney Yorke, a commissioner in the British Admiralty Office and a Rear Admiral in the navy. How the island became known as Bostwick has been the subject of conjecture over the years. One version has it that Bostwick is adapted from Boss Dick Island, a name given locally in reference to a man who ran a granite quarry there. Another version is that a person named Bostwick was one of the first guests to stay at the Bostwick Island Guest House, run by the island's original settler-farmer family, the Turcotts.

Collice Turcott first settled on the island in 1857, having purchased it from the widow of the Honourable John McDonald, who had himself purchased this island's lease and several others from the Mississauga Indians, for the sum of $100. As you cruise along about the halfway point on the north shore of the island, you can glimpse the remains of the Turcott farm at the back of the bay behind the cottage called Isle of View. Turcott had built a log house and barn and planted an orchard in the central part of the island. Because of some problems in the 1870s when the Department of Indian Affairs would not honor the original Indian lease, Turcott was forced to buy the island again, which he did through a loan from a Kingston merchant, James Richmond. Over the years, the Turcotts sold cottage lots on the island and rebuilt the farm house into a commercial guest house.

In 1882, John Wallace, a Bostonian, bought properties on Bostwick, including Isle of View, the cottage called Iroquois, which he renovated with the towers seen on it today, and the land that surrounds Half Moon Bay. Since 1887, church services have been held in the open air cathedral of Half Moon Bay every summer Sunday, enjoyed by all who live in or visit the Thousand Islands.

The bath house at Iroquois Cottage on Bostwick Island.

A TALE OF HALF MOON BAY

The preacher's voice resonated from the rock walls of Half Moon Bay, carrying easily over the assembled fleet of skiffs, canoes, and small steam launches. Inside the remarkably curved arms of the bay, a shape which prompted the name for the place, the minister from Gananoque took his turn at the stone-hewn pulpit. Since every denomination in the town was given the opportunity to hold services in the bay just once each summer, he was going to relish every moment of the experience. After all, attending a service at Half Moon Bay was much more a rendezvous than an obligation. The outing to church was often just the beginning of a social day: picnic baskets and fishing rods were nestled under the seats, and invitations to afternoon island teas were exchanged.

Islanders, townspeople, and visitors alike found their way to the south side of Bostwick Island, not far off Wanderer's Channel. The boats eased into the bay, wedging and interlocking like puzzle pieces, bow along stern and gunnel to gunnel. Those closest to the rock wall looped lines over the rod wire that runs conveniently along most of the length of the bay. Others traded bow or stern lines or simply reached out to hold a neighboring craft, the whole effect as if everyone was joined together in the spirit of the place. Being a fine August Sunday, the fleet of churchgoers was particularly large. And being past the mosquito season, the assemblage was quite at ease, allowing the minister to settle comfortably into his sermon.

What an inspiring place is Half Moon Bay! Could there be a better place to rally the human spirit! This open-air church is like no other. A pulpit of a granite block stands on a rock shelf near the back of the bay. Arches of maple and ash bows cast graceful shade over the shores, but the true ceiling of the church is the sky overhead. On the north shore of the bay, the green woodland cloaks a sloping shore that dips gently into the water. The south curve of the bay is much more dramatic: a superb backdrop for the atmosphere of a church. A bluff guards the south entrance to the bay and the high ground continues along the bay itself. One of the more unusual geological features of the Thousand Islands is found where the rock walls pinch together at the back of the bay, a series of huge cauldron-like pits, called potholes, in the granite. Some are big enough to hold a half dozen people. Others are broad but shallow. After seeing the big potholes, the origin of the curved rock surface of the south side of the bay is clear. The unique shape of the bay formed when the gravel-laden waters of melting glaciers swirled and poured over the island's surface, wearing holes into the rock. The fluted patterns in the rock wall may be from where the sides of some potholes wore through the cliff face, and from the rush of water through the back of the bay.

Postcard of a summer service at Half Moon Bay.

Services have been held at Half Moon Bay every Sunday in July and August, weather permitting, since 1887. The inspiration for this open-air cathedral came when a boating party took thankful refuge in the bay during one of those sudden and fierce summer squalls on the river. The land around the bay was purchased in 1901 by a Bostonian named David Wallace. When he died in 1904, the land was willed to the town of Gananoque so that the church would always continue.

TREMONT PARK ISLAND

Tremont Park Island, called Tidds Island on some charts, lies just south of the town of Gananoque. Originally the island was named for a squatter, William Tidds, who lived and farmed on the island in the 1820s. This relatively level, oval-shaped island has very likely attracted people for the summer months long before recorded history. In the late 1800s, a native burial site was found there. Sheltered by its neighboring islands, Tidds offers a great view up and down the river and there is excellent fishing in the shallow waters to the south and east — all qualities enjoyed by peoples throughout time in the Thousand Islands. No doubt, all of these things prompted Captain Sanford Davis to purchase the island in 1878, where he founded a summer community there called Tremont Park. Davis divided the east end of Tidds Island into cottage lots, selling them on the condition that no liquor could be served. A boarding house was built at the western end of the island which became popular because it was easily accessible from the town and because of the tours that were run so patrons could enjoy the river scene. As the years went by, the boarding house was expanded and adapted as a hotel. Alas, it was destroyed in a fire in the 1920s, but many of the other cottages from those early days still stand. A slow tour around Park Tremont Island to view the character of the old summer homes is a pleasant excursion.

The boathouse on Niagara Island.

NIAGARA ISLAND

One of the most striking cottages in the Islands looks like it was transplanted from the Mediterranean — and in a sense, it was. Sherman Pratt bought Niagara Island in 1929 and soon after, decided to build a cottage. The story goes that Pratt asked an architect friend, Jack Woods, to design and construct a place for him. When Pratt arrived in the summer of 1931 to take up residence, he discovered that Woods had decided to use the same plans as he'd produced for a home in the Mediterranean. The Art Deco style and concrete construction is unlike anything else in the region and always surprises tourists who 'discover' it each season. The sandy tan-colored building has been well maintained and remains in the same family. A substantial part of Niagara Island remains in a natural state, from forest floor to treetop, and so these landowners also contribute to protecting the ecology of the Thousand Islands.

HOUSE OF SEVEN GABLES

About three islands west of the Canadian span of the Thousand Islands International Bridge is an island variously known as Pine, Dashwood, or Hime's Island. The impressive summer home on the south side of the island facing Hill Island features in the commentary of every tourboat that travels this section of the river. Built sometime just after 1900 by a gentleman named Davis, possibly from Ottawa, this cottage is locally known as the 'House of Seven Gables'. The historical name, Opawaka Lodge, apparently meaning 'swift water', was given by the island's second owner, Joseph Himes, who bought it for $16,000 in 1923. Himes served a term in the U.S. House of Representatives but is often referred to as a Senator in some tour discourses.

Pump house and docks at House of Seven Gables.

The buildings and landscaping reflect the ideals of an island owner at that time. Included among the buildings are a water tower and a pumphouse, boathouses and docks, a powerhouse and an icehouse, and a 'proper' flagpole. While the powerhouse and icehouse were important for island life in their day, electricity has long since made them obsolete. They now serve as tool sheds and storage buildings. Another character of the island harkening back to a century-old vision of an ideal island life is the way that the grounds are kept. Even from the water, it's clear that the undergrowth has always been 'kept down', showing human dominance over the natural wildness of the land.

SURVEYOR'S ISLAND

Tiny Surveyor's Island was named for a Mr Rubidge, a Dominion surveyor, who camped there in the 1880s while making a survey of native lands. Because the island is in the main channel of the river a little east of the Canadian span of Thousand Islands International Bridge, everyone who passes by has a chance to view the island — and vice versa. Those who take a closer look notice a place that is the very picture of a Victorian summer cottage. Dusty dark green and deep red colors complement the intricate lines of the building. Imagine what the builders would have said, back in the early 1900s, if they had been told there would be an immense steel suspension bridge, carrying over two million vehicles per year, framing the sunset view from Surveyor's Island.

Surveyor's Island at the Thousand Islands International Bridge.

DARLINGSIDE

The 'Canadian Palisades' are high granite cliffs along the north shore of the river between the Thousand Islands International Bridge and Rockport. These cliffs drop straight into the river, and the charts show depths here reaching the better part of 100 meters. At the west end of the cliffs, there is a fairly level stretch of ground, and in this dramatic setting stands one of the last surviving steamer fuel-wood depots along the Great Lakes. Darlingside is named for its founder, Thomas Darling, who emigrated from Scotland in 1837 and started a river-front store on this site soon after. As it turned out, both the location and the timing were fortunate.

A valley runs inland through the intimidating granite cliffs of the Palisades, linking the store to farms inland, the village of Lansdowne, and tracts of timber.

During the early 1800s the river remained the best corridor for travel since roads were still little more than cart paths, passable only in the dry months of summer and fall. While sailing ships would still be built until the 1890s to haul cargo on the Great Lakes, steam power was relied upon to overcome the rapids and currents of the St. Lawrence. Some work had been done on canals by the late 1830s to bypass the rapids between Prescott and Montreal, and steamships were becoming increasingly more reliable and powerful, with a voracious appetite for fuel. Thomas Darling, like many others along the river, seized the opportunity to provide a supply. Since coal didn't replace wood as fuel until the 1870s, Darling's depot was able to carry on for several decades.

The store at Darlingside, the oldest fuelwood depot still standing.

There is 15 to 20 meters of water immediately offshore from Darlingside, so bringing the freighters right to the shoreline wharf was not a problem. Although all that remains today are the big iron mooring rings that were drilled into the granite, there was at one time a big dock on piles at the edge of the river. Steamers would put alongside to load cordwood, cheese, and farm produce, and to unload supplies and special orders for the Darlingside store. By the late 1840s, the store was a going concern and carried everything from Darling's specially branded tea to harness. Darling was even an agent for several Montreal wholesalers. As the need for fuel grew, Darling had to go ever-further inland for wood. Farmers for miles inland bartered wood for credit at the store, and in this way, the cordwood business at the store became a very important part of the local economy. In fact, when rail lines were built through the village of Lansdowne, it was only natural that Darling would set up stores there, too, and continue the growth of his trade.

The Darlingside store stands just back from the water's edge, a well-built and well-proportioned Classical Revival style structure, which suggests that Darling knew he would be in business here for some time. The house, just west of the store, is actually two separate buildings that have been put together in a very aesthetically pleasing way with clapboard siding and ornate woodwork. Outwardly, there has been little change in the store and house since the 1880s. The Darlingside store, in private ownership today, has been given National Historic Site status because it is the last of the steamer fuel-wood depots and stores still standing on the St. Lawrence River.

ROCKPORT

Andrew Cornwall, an Alexandria Bay businessman, was one of the principal developers in the Thousand Islands region in the late 1800s. Businessmen would often have ventures on both sides of the river in those days. In addition to his many real estate holdings and his store on the American side, Cornwall built a store at Stony Point, now the Canadian village of Rockport. Andrew's brother Charles came with his wife to run the enterprise, which sold fuel wood to steamers and general supplies to the rural community. Charles eventually bought the store around 1850, and in 1866, built a big house beside it — but in back of the big rock that hid the river from his sister-in-law's view. She ordered the rock to be blasted away. After Mrs Cornwall had the promontory blasted off to improve her view of the river, it didn't make much sense to call the place Stony Point anymore. The name Rockport seemed appropriate. Ships sought shelter in the bay at the east edge of the village during especially fierce blows and the village was a port serving the neighboring farm community.

The operation remained in Cornwall hands until 1936. Today, the store is the Boathouse Restaurant and the old house beside it operates as a tavern.

CLAYTON

Clayton was a booming summer resort in the 1890s. The Thousand Islands had become so popular that Clayton had, in effect, become the U.S. gateway to the region, with up to 13 trains arriving daily from New York, Syracuse, Albany, and other centers. Many steamers, including the 1,000 passenger *Empire State*, met the trains and carried travelers to hotels and resorts throughout the region. This was the place to see and be seen in.

In 1881, Charles Emery, founder of Brooklyn's Goodwin and Company tobacco firm, bought a small group of islands off Clayton. He renamed Powderhorn, the largest of these, as Calumet, meaning Indian Pipe of Peace. The original shape of the island apparently resembled a native peace pipe. Calumet is the most prominent island seen from Clayton, and Emery took great advantage of that, creating a showplace by reshaping and linking the island cluster to create a lagoon that would shelter large and small boats. Several other buildings were constructed, including a water tower, guest house, skiff house, ice house, and his summer residence. The main frame building was soon replaced with a much bigger one of stone, quarried from nearby Picton Island. An unusual feature of the new house was a cyclone cellar, insisted upon by Mrs. Emery, who was perhaps intimidated by the brisk afternoon winds on this section of the river, even though there have never been cyclones here.

Emery became very involved in the summer life on the river. His island gained notoriety, with frameworks set up on the shore from which lanterns could be hung to spell out various names, events, and shapes. There were fireworks on the island every summer evening at eight o'clock. While Emery wasn't the first to have lights and lanterns, his setting was one of the most prominent.

As well, Emery bought the Frontenac Hotel on Round Island, just off the east edge of the Clayton waterfront, and turned it into the social hub of the area. Many of the hotels of the time restricted their events to their patrons, but this wasn't so at the Frontenac. Prominent islanders and village guests came by boat to the hotel

The Charles Emery house on Calumet Island off Clayton.
(The Thousand Islands, James Bayne Company)

for evening dances and parties that would often last throughout the night. Colorful lanterns decorated the lawns. The big dining room was decorated with boughs of evergreens, cut flowers, and flags. Guests, in dress suits and silk dresses, danced, socialized, and chatted the nights away. Hotel guests themselves had endless choices of things to do. There were fishing guides on hand, skiff liveries, steamer tours night and day, a nine-hole golf course, children's activities including pony rides, and much more. The Frontenac set the pace and other hotels followed suit, making summer in the Thousand Islands an event as well as an experience.

Unfortunately, the Emery castle on Calumet burned to the ground in the mid 1950s, after having been abandoned for several years. The water tower and the majority of the other buildings still stand, hinting at the wealth of the island's original owner.

MURDER ON MAPLE ISLAND

For the citizens of Clayton, the incident had all the ingredients of a good mystery novel. It was the early part of the month of June 1865, and there were as yet few people living in the region, let alone spending summers on islands. A stranger was sure to be noticed right away. The man rowed over from Gananoque in a skiff and took a room at a hotel at Fisher's Landing. He spent a few days exploring and fishing, keeping pretty much to himself. Recalled one local man in the sleuthing of the events that were to follow, "He was about 30 years of age, with black hair, eyes and beard, well dressed, very uncommunicative, dark as a Spaniard, and very restless."

No doubt, there were some who warmed to the stranger when he employed a few carpenters to help with putting up a cottage on Maple Island, a little to the north and east of the village of Clayton. The cottage was built on a bluff that would have given it a good view over the river, but it was screened from view from the water by bushes. The work was done in short order, and again the man kept to himself, with just his books and violin for company.

One night, there was an orange glow across the water over the island. People in the area assumed there was a fire, but figured the man would have escaped and he would show up at the village the next morning. When he didn't arrive, a party went out to see what had happened. What they did find set the whole village to talking. The man had been murdered. His throat had been slashed and there were three cross-shaped knife cuts in a triangular pattern on his chest.

Now as it happened, a week before the murder several men, assumed to be Southerners by their accents, had been seen around various hotels in Clayton. Interestingly enough, they had set out by skiff, supposedly for Alexandria Bay, the evening of the murder. The cuts on the dead man were recognized as a sign for the secret society, the Knights of the Golden Circle. The most popular theory floated in the Islands was that the stranger was none other than the treasurer of the society, a man named John A. Payne, who had made off with $100,000 of the blood money paid to the society for the assassination of President Lincoln. It appeared that Payne had been hunted down and killed for running out on the society. The murder was never solved and exactly what transpired that night on Maple Island will never be known.

This story was recorded in *The Picturesque St. Lawrence*, written as a souvenir of trips to the Thousand Islands by J.A. Haddock in 1896.

GANANOQUE CANOE CLUB

Vacationers and residents of the Thousand Islands have had the joy of racing just about every type of craft there is on these protected waters. Among the most enduring of the events are canoeing and rowing.

Canoe racing began on the Gananoque section of the river in the 1880s. Paddling events, skiff racing, and sailing canoes were all part of the fun. Summer afternoons on the river can be quite blustery, and sailing canoes, with huge spreads of light canvas and only the weight of the sailor for ballast and steering, called for great skill. The canoes flashed over the water and around the buoys with plenty of dunkings along the way.

The American Canoe Association races held at
Grindstone Island were major events in their day.
(Antique Boat Museum)

The American Canoe Association met on Canoe Point on Grindstone Island in 1884 and 1885, and again on Stave Island off Landons Bay in 1889. Soon afterward, the Association purchased Sugar Island and still owns that island today.

The town of Gananoque was ready to have its own club by 1906, and a clubhouse was built on the waterfront in 1908. The Gananoque Canoe Club became a social centerpiece for the town. At the time, the upper floor boasted the biggest dance floor in eastern Ontario. Today, the canoe club still operates from the same building, with all of its turn-of-the-century charm. It is now owned by the Rotary Club and hosts a wonderful summer theater, called the Gananoque Play House.

Annual meeting of the American Canoe Association, Hay Island.
(The Thousand Islands, *James Bayne Company*)

Skiffs & Runabouts

MESSING ABOUT IN BOATS

The Nina in the
Thousand Islands.

After some 15 years of guiding, the act of pulling the skiff alongside the dock and flipping a line over a cleat was as smooth and natural as breathing. The rods were ready, the bait was sloshing in the tin box under the centre seat, and the cane chair in the stern was ready for the sportsman of the day ... This was going to be a most beautiful day on the river. The last of the bright stars was just now fading into the vanishing night, and the eastern sky was becoming the hammered and weathered silver color of the river guide's favorite bass spoon. A glance over his shoulder confirmed that the dock of the Grenadier Island Hotel was not far off. The row down from Rockport seemed to have taken no time at all. There was an orange glow from nearly every room at the inn, but the brightest pooled out onto to the lawn from the dining room. It seemed like all of the guests realized that in the first weeks of June, the bass would almost seem eager to be caught. None of the sports was going to miss out on the fishing — or more likely the opportunity to brag about the length of the stringer of fish they would bring back to the hotel. Used to be that they just came for the fishing, the solitude of the river, and the incomparable shore dinner. Then one day someone came up with the idea of taking pictures of the fisherman, or perhaps his wife — usually grim faced, standing next to a *pike swinging in the breeze — with the catch of the day. That was when it came to be that the score for the day was as important as the event of fishing itself.*

Just when he was ready to stroll up to the inn to see who was his own catch of the day, he heard the cough then wheeze then the metallic, thudding roar of the motor. So, they'd found this place, too. Years back, there'd been only the guide and the skiff and his clients. Then had come steam engines small enough to be put into stubby little boats that could tow a string of skiffs and their guides and sportsmen out to the fishing grounds. That hadn't been so bad — at least when they'd wanted, the guide could cast himself loose and be alone at his trade. But now the gasoline engine had come along and this younger generation of river rats was giving their clientele the fine notion that they could fish just about every part of the river. Now they could travel upcurrent and upwind any distance they chose, speeding those city folk to all the best shoals and bays. Those quiet conversations where just about every topic under the sun was brought out for discussion? Gone. Replaced by whoops as spray flew back over the bow and any semblance of peace was lost to the yammering of that bellowing engine. Well, at least he still knew of special shallows and shores those newcomers couldn't get to. There was still a group who would pay to appreciate the fine points of fishing in the Thousand Islands.

Watercraft of some sort have plied the waters of the Thousand Islands since man first came to the region. Boats of one kind or another figure in the discovery, exploration, commerce, military history, and, perhaps above all, the enjoyment of the region. Boat building materials were close at hand in the surrounding forests. Log rafts and dugout canoes transported the first peoples to the Islands as far back as 9,000 years ago. When European explorers arrived to search for a 'Northwest Passage' through the heart of the continent, they adopted the native canoe, which proved to be just the thing for traveling quickly on the rivers and safely through the rapids. As the need arose to carry more cargo and passengers, European traders developed the delicate bark canoes into much larger freight canoes. Before long, bigger and heavier rowing boats — bateaux and Durham boats — took expeditions upriver and furs down. Border wars and skirmishes between the British and the Americans prompted the construction of navy vessels.

For the first settlers of the Thousand Islands, boats provided their only link to the outside world. Most every community had boatwrights and shipyards. Boat building

took place chiefly in the fall and winter, after the harvests were in for the year. Farmers and farmhands were resourceful people who could turn a hand to the special construction techniques required. All sorts of craft were needed in each community. Rowboats, workboats, scows, and freight carriers each had their place. Small maneuverable boats were needed to travel among the islands. Larger ships were needed to carry the efforts of trade and commerce through the Islands, from the ocean into the continent and back again. Builders came to specialize in one general type of craft or another. Ideas for shapes and lines were freely borrowed from other boatwrights throughout the region. There were no standard plans for their boats at first — practically every boat built was unique — but eventually designs evolved that were best suited to the Thousand Islands.

President Chester A. Arthur (top) enjoyed fishing from the skiffs. (Antique Boat Museum) *St. Lawrence skiff at Calumet Island.*

THE ST. LAWRENCE SKIFF

The myriad of twisting channels, shoals, and strong currents demanded special designs, and although there are many sheltered bays and channels in the Thousand Islands, the open river is often less than tranquil. There is a lot of big water here and winds and storms often rage along its course. Boats cannot be frail, yet at the same time they must be relatively shallow in draft and very maneuverable to thread the intricacy of the isles.

Sometime in the mid 1800s one of the first important Thousand Island boat designs appeared in the construction of what became known as the St. Lawrence skiff. If indeed form follows function, then it was natural that the St. Lawrence skiff would be developed by local craftsmen to suit this region. Double-ended rowing boats were familiar to settlers from Europe and Britain, but none of these made perfect sense in the

Thousand Islands. Some were too light. Some were too cumbersome for the distances and conditions here. The 'lapstrake' building method of overlapping plank upon plank was also a well-established technique that had been in use even in the days of Viking longships. The St. Lawrence skiff merged and modified the best features of these craft into a unique design.

Although the skiffs ranged in length from about five to six meters (16 to 20 feet), most were about 5.5 meters (18 feet) long with a beam of about 1.04 meters (40 to 42 inches). Almost all had seven strakes, or overlapping boards, per side — the exception being the Andress' skiffs from Rockport, which had six — and were built of white cedar or sometimes white pine, with fine ribs and keels of white oak. Skiffs were pointed at both ends but were not quite symmetrical. They have a shape that is a little fuller towards the stern, so that if there is just one passenger, that person can sit at the stern to face the oarsman without having the stern sink so much as to hinder the performance of the craft. The bow and stern rise much higher than the low freeboard so that the skiff can ride dry through the chop on the river but be rowed efficiently. There is a short deck at both bow and stern to stiffen the craft and help to keep it dry in rough water. Oars were generally 2.3 meters (7 1/2 feet) long, and often had cupped blades. The oars were mounted on thole pins rather than oarlocks, and many skiffs had two sets of pins so that the skiff could be rowed well balanced, depending on whether the oarsman was alone or had cargo or crew.

In the late 19th century, St. Lawrence skiffs played a major role in the early guiding and tourist business on the river. Several builders turned out their versions of the skiff, and the character of these ranged from a workhorse to a work of art. Those built for fishing guides often had special fittings, including rod holders and a tin-lined drawer that could hold water, slid under the seat to hold bait and, hopefully, the catch. St. Lawrence skiffs built for more well-to-do owners often had special trim of exotic woods, wicker-back seats, nickel-plated brass fittings, and flawless varnished finishes.

Rowing is not only an opportunity to think about where you are going, but also about where you've been. Perhaps it has something to do with looking over the transom as opposed to over the bow. When a craft rows as easily as a St. Lawrence skiff,

A fishing expedition shore dinner (top) and the day's catch, late 1800s. (Antique Boat Museum)

there is not only a chance to think but the opportunity to reflect. A boat such as this glides through the water rather than pushing the river out of the way. Yards rather than inches yield to the oars, and the skiff's course is straight and true. A quiet 'V' of wake streams away from the pointed stern as the river whispers along the elegant lines of the full and rounded sides. From time to time, the water chuckles as it's turned by the honey-colored overlapping cedar planks of the skiff's flanks. If ever there was poetry in motion, it would be the St. Lawrence skiff.

Although a great many of the skiffs have been lost to neglect and time, there are still several in good shape to this very day, kept in boathouses among the islands. These are enjoyed to the fullest by descendants of the original owners and others who simply love the way these incomparable craft slip through the water.

Fishing parties (top) were often towed to the "hot spots" by steamers. (The Thousand Islands, James Bayne Company).

Postcard of the Riot, *one of the speedboats that could be hired for thrill rides in the early 20th century.*

RUNABOUTS

In the years immediately after the turn of the 19th century when the Thousand Islands boomed as a summer place, boatbuilders had plenty of work. Ferries, steam-powered tourboats, work boats, and barges were needed to move people and materials. As engine technology improved, mainlanders and islanders who had the money wanted launches and commuter boats built for their needs, and when gasoline engines came on the scene, it didn't take boat builders long to adapt them for use on the water. Some historians have claimed that the fast gas-powered runabout was first developed in the Thousand Islands. Newspapers of the day were certainly full of accounts of the fast boats and the people who raced them here. Among Island runabout builders, the Hutchinson Brothers, Gilbert, Fay and Bowen, and Stanley were the most famous. The quality and variety of their work rivaled the craft from

more widely known national builders, such as Chris Craft, Gar Wood, and Hacker, whose boats were brought to the Thousand Islands.

As in the case of the skiff, the runabouts were built to reflect the demands of the region. Speed, of course, was an obvious requirement for any runabout. Considerable trial and error went into finding hull shapes that would plane well yet ride the choppy water in comfort. The most glamorous boats built were the double and triple cockpit runabouts, built for speed and fun. However, the majority of the boats reflected the variety of needs of summer island and mainland residents. The boats needed to carry supplies and people back and forth along the river, all the while accommodating that favorite activity, fishing. The resulting designs were fast boats with open space around the engine box and relatively low freeboard towards the stern to make it easy to get in and out at the dock and to make it easy to fish. The boats had to be big enough for the rough water on the river, but small enough to handle well at little docks in protected bays. Since wooden boats are best kept in boathouses, they had to be small and light enough to be hauled out on their lifting eyes in the fall.

There was another breed of powerboat built to evade the law invoked by the 18th amendment to the U.S. Constitution, better known as Prohibition. With that law, in force from 1920 to 1933, it was illegal to import liquor into the United States. Hollywood has made fast cars and powerful boats a glamorous part of the lore of that period. Even today when some see wooden boats roar by, they excitedly point out the 'rum-runners' to their friends. Boats were in fact specially built for night-time cross-border cargo transfer. Some had brawny engines, even engines modified from aircraft, to outrun the best the Canadian or American governments could put to the chase. There are stories of builders having two boats under construction at the same time: one for the 'trade' and one that would probably wind up in its pursuit. Contrary to the movie ideal, however, the majority of river crossings with cargoes of amber gold were made in much less spectacular craft. Skiffs, duck

A St. Lawrence skiff (top) and the annual Antique Boat Show at Clayton's Antique Boat Museum.

punts, and old farm work boats were a lot less likely to draw the attention of the revenue men.

There are still many wooden runabouts in use in the Thousand Islands. They are tucked into boathouses, out of the harsh sun to protect their bright varnished hulls and decks and to keep the upholstery out of the weather. On fine summer days, in the flat water of early evening, their owners take sojourns among the isles, with the lusty engines echoing throaty rumbles from the rock faces of the island shores.

One of the best places in the world to see wooden boats is the Antique Shipyard Boat Museum in the village of Clayton, New York. The Museum houses one of the world's finest collections of skiffs, canoes, raceboats, and runabouts. It's open through spring, summer, and fall months so that visitors can see restored craft in several buildings and floating in their slips. Rides can be arranged in some of the powerboats. Ongoing expansion will let people explore an ever-increasing collection that will include George Boldt's luxurious turn-of-the-century houseboat, La Duchesse, which has floated in a sheltered bay just west of the castle for many years as the summer home of Andrew McNally III; and a number of powerful wood-hulled raceboats, some of which roared over these river waters. Visitors will also be able to take a St. Lawrence skiff, which may have been faithfully reproduced at the Museum's workshop, for a delightful row.

Two boathouses at the east end of Ash Island.

BOATHOUSES

Boathouses, too, are an integral part of the traditional scene along the river. No two are the same. Some of the earliest boathouses were quite small and actually sat on the shore. They had ramps down to the water so that the skiffs and canoes they housed could be slid into the river when the moment came for a tour. When powerboats came on the scene, boathouses were built on cribs, which are like oversize 'crates' of heavy timbers filled with rock, set in the water along sheltered shorelines. Some boathouses are modest affairs, little more than garages for boats. Others are bigger and more elaborate, housing two or more boats and all the gear that might go with vacations on the river.

There are boathouses with cottages or guest accommodation over the slips, often decked out with verandas and gingerbread trim. A few were very large, on the same scale as the castle-like summer homes they served, housing the yachts of their wealthy owners.

In the boathouse, the world outside seems distant, muffled. All sounds and footfalls have a hollow, dusky tone. There are sounds that could only come from such a place — a soft-padded splash from water turned away from the gently rocking boat hull, a lazy gurgle of water through the algae-softened timbers of the cribs, the tiny protest of rope tugged against a mooring cleat. Faint scents can be smelled: a slight mustiness of old wood and cobweb, the wet and dank — almost earthy — smell of waterweed and algae exposed on the waterline timbers, a whiff of oil that has found its way to the boat's bilge, and the fragrance of sun-warmed pine and blueberry heath that strays in from the world outside. Shafts of white sunlight stab into the dusty shadows, highlighting the planks of the walkways and weathered boards of the interior walls. The emerald green river water glows where sunlight slants through the open door and streaks through the vertical dock staves that dip like dozens of straight fingers into the water. Wavelets are like constantly rippling lenses that focus flickering dapples of sunlight onto the silty bottom of the boathouse, brightening into view for an instant a waterlogged twig, a clam shell, or a glass bottle accidentally dropped years ago. Some of these same ripples reflect dancing crescents of light onto the dark interior walls. If this is a particularly lucky boathouse, the gleaming red-brown and caramel hull of a mahogany runabout rests here out of the sun and seems to float as if suspended on the surface of the clear green-hued water.

When people enter this little realm, adjusting their eyes to the dim light, there is a quick flurry of activity from the inhabitants. If a family of barn swallows has nested on a rafter, they take wing out the open front of the boathouse, screeching shrill double chirps of warnings and indignation, darting in and out, over and over again. The store-bought, pressed-paper model of a great horned owl that was to discourage the swallows hangs by a piece of cord in the center of the building, seemingly connected as much by a cobweb as by a cord to the overhead beam. Spiders, alert to the possibilities of an insect lunch bumbling onto their webs stretched between beam and post or among old coils of

Postcard of the Boldt Castle boathouse, with La Duchesse *and the yacht* Clover *in front, and room for the rest of his 18 boat fleet inside.*

rope, sense a person's movements and vanish into shadows. Perch and sunfish, lounging in the cool shadows of the boathouse, drift warily into the deeper shadows of the cribs or under the boat. The residents will again take up their accustomed haunts once the mission to the boathouse is accomplished, be it so simple as checking the lines on the boat or the noisier, water-swirling commotion of taking it out for a tour of the river.

These boathouses are cool, quiet havens, places where everything nautical about the cottage or summer home is kept out of the weather and ready for use. On the walls are hung all sorts of things for boat and river: oars, paddles, life jackets, boathooks, short and long lengths of rope, along with webbing and chains for hauling boats. There are often clusters of fishing rods and nets standing in corners, small boat spars and sails up in the rafters, and all sorts of tools and parts for motors, some of which haven't been around the old cottage for years. The motto for organizing the boathouse might be, "a place for everything, and everything all over the place."

TOURBOATS

I f it weren't for the tourboats, few of us would have a chance to see the Thousand Islands from the water. Providing tours of the region by boat has been a business here practically from the day that people were drawn to see the beauty of the river and the islands. Views from the shores are terrific, but seeing the islands up close is a memorable experience.

Boat tours have been a competitive business right from the start. When Captain Elisha W. Visger's small steamer, the *Cygnet*, was launched in the mid-1870s to begin the first scheduled tours on the river, others were quick to see the potential, and boatlines began operating from Alexandria Bay, Clayton, Kingston, and Gananoque, much as they do today. To attract customers, the boats boasted of their furnishings and comforts. Many of the boats had elaborate and ornate woodwork, rich carpets, sunshades and awnings that were as stately as some of the era's grand hotels. Captains chose their routes with care to offer the best of views. Some added night cruises to their schedules,

TOURBOATS

illuminating cliffs and cottages with powerful spotlights while bands played for the entertainment of the guests. In fact, the night cruises were such an event that many of the islanders lit an array of colored lanterns to the delight of the tourists as well as their own guests.

The First World War and Great Depression in the first decades of the 20th century dealt hard blows to the tourism business of the Thousand Islands. During those same years, automobiles became reliable enough to travel considerable distances. People could now visit the Thousand Islands without having to take long trips by train and could stay at any place they chose for as little or as much time as they wanted. Tourists were no longer confined to the towns where the trains ran and were no longer delivered en masse to the waterfronts where tourboats waited.

Between the evolution of travel and the changes in the technology in engines, the end was quickly in sight for the river's old steamers. Tourboats became smaller and ran shorter tours. These new tourboats were single-decked, wood-hulled vessels powered by big gasoline engines. Glass windows all around offered a good view and a coach roof overhead kept out the weather. While these types of boats lacked the grand opulence and studied comfort of the steam tourboats, they did offer regular and frequent schedules and even a little of that feeling of being in a powerboat. Speedboat rides in the Islands were also available to tourists for a few years. Mahogany-hulled craft were driven to speeds of 40 and 50 miles per hour by converted aircraft engines. Thrilled passengers hung on and shouted over the thunderous roar as the boats sped through channels, with curtains of spray hurled away from— or sometimes wind-thrown back over — the passengers.

A *tourboat carries picnicers to Beau Rivage Island, about 1930.*
(National Archives of Canada C20527)

The steamers Brockville and St. Lawrence were two of the best known on the river. (The Thousand Islands, James Bayne Company)

Sport (opposite) was a private yacht serving Sport Island near Alexandria Bay. (Parks Canada)

When the numbers of tourists began to increase again in the 1960s and '70s, the need for larger boats returned. Double and then triple-decked boats with designs that mimicked everything from Mississippi River boats to oversized speedboats were introduced into the tour trade. There are only a few of the little wood-hulled boats left plying the river. History seemed almost to repeat itself: tourists were brought to the region en masse, but this time it was buses rather than trains that were the carriers. These visitors are likely to see the islands by prearranged, packaged tours. But for those who have the freedom of choice, there are still a wonderful array of options. Since the various lines begin their trips from different points on the river, taking more than one tour is a way to see other parts of the Thousand Islands. For those who want customize their tour of the Islands, there are even water taxis for charter, and several touring kayak operations can supply both boats and guides. In addition, many marinas rent smallboats, mainly outboards, by the day or week. While advertising seems to suggest that Boldt Castle is the highlight destination, consider any of the tours as a voyage of discovery. There is plenty more out there than the castle of one man's dreams.

Channels & Bridges

ENTERING THE MAZE

Island passages.

The night started out quiet enough. The two customs agents had been invited to the island near Ivy Lea for a little poker. Some money was on the table that night and, funny thing, just like the last couple of times that they'd been invited to play, one of them was winning. These locals, it seemed, just didn't have a real sense of the game. A bit of bluffing, a couple of clever moves and you could walk away with more cash than you'd make in a month of wages as a law man. Then, right about when they'd nearly cleaned their hosts out, they'd heard the powerboat roar by. The time of night, the direction the boat was headed — it would have to be smugglers. They'd grabbed their winnings and ran down the dock to their own boat, but the lead was too great. All they heard was the wash of boat wake along the night-black shores and the echo of a speedboat's engine somewhere to the west. The smugglers had vanished upriver, taking the twisting channels past Pine and Wallace Islands at breakneck speed. By now they were likely unloading in some shadowed bay on the American shore. All the searching you like will never take you into every channel and bay in this part of the river . . . and it was highly unlikely anyone would come forward and volunteer information about any illegitimate boat crossing the river that night.

Whether you travel the river by tourboat or on a boat of your own, your passage is bound to take you along some of the memorable channels of the Thousand Islands. Wherever there are clusters of islands, the river narrows and twists its flow to the whim of the irregular rocky shores. The channels are sometimes broad avenues through the isles and sometimes quiet and mysterious laneways that weave among rocks and islands, big and small. Some beckon boaters; others will accommodate only skiff or canoe. All are unique and have their own stories to tell.

Many of the channels through the Islands are named on the charts. Most of these have been used since time first remembered because they are the most logical and safest routes among the islands. Names in some cases harken to the role that the channel played in the region's developing story; others are simply popularly accepted local names, now part of the tradition of the Thousand Islands.

LIGHTING THE WAY

"Red, right, return": keep the red channel markers on your right hand when your craft is headed towards the center of the continent. That sounds like a simple enough rule for basic navigation — until you see any of the nautical charts for the Thousand Islands. The passages through the islands are all over the map, so to speak. Some are obviously main channels, but dozens more are side and connecting channels. Even with charts as good as they are today, and with spars and buoys set in the major channels to guide the way, there are a lot of sorry looking bent propellers as souvenirs of summer visits.

The problem with navigating through the Thousand Islands is not the islands themselves, but the rocks and shoals that lurk beneath the water's surface. The Thousand Islands are a flooded landscape of hills, cliffs, valleys, and canyons. The lay of the underwater landscape is only a little predictable. The orientation of the hills and valleys is slightly diagonally across the river: this is the trend of the rolling topography of the foundation geology, the roots of an ancient mountain chain. The odds are just a little better that one will find shoals to the east and west of islands than finding them to the north of south, but don't count on it.

When the Thousand Islands were first being settled, small boats were the only vessels that could mount the rapids above Montreal and avoid the visible shoals. When ships became larger and steam-powered in the 1830s and '40s, the shoals in the river became a true menace. People on the shore were no less distressed than passengers and crew on board to see the ships that carried their supplies or the products of their hard labor run aground and even sink, often within sight of their town docks. River captains in those days had a very good knowledge of the river, but sometimes, with storms, fog, or mechanical problems, things could go wrong.

Not until 1856 did the government place markers at key points in the Islands. Before that time, local residents often took matters into their own hands. Such was the case to the east of Gananoque at a rocky shallows where there are no easy landmarks to locate the shoals. In 1840, James Parmenter, or 'Jack' as his friends knew him, put an earth-filled barrel on the rocks, which barely broke the surface. The barrel noted where danger lay,

The tourboat New Adonis *heading downstream past the Little Lyndoch Light.*

but the marker became even more noticeable when the weed and grass seeds sprouted in the barrel's wet soil. That prompted the locals to name the place 'Jackstraw Shoal'. Because waves were not long in washing the cask away, a stone-filled barrel with a bush in it was set on the rocks, but this too had to be replaced from time to time. A government marker was set there in 1856, the same year that markers were placed at various locations in the Admiralty group of islands. A light and beacon were put out to mark the shoal in 1880, but these, too, have gone. Today, a steel cylinder and light show the north side of the channel and a floating green buoy marks the edge of the rocks to the south.

By the early 1900s, several of the river's most noticeable navigation trouble spots were marked by lighthouses. These provided seasonal work for a few hardy rivermen of the day. One such post was the Gananoque Narrows, a tricky channel between the head of Stave Island and the nearby foot of Prince Regent Island. A boat must pass deceptively close to the shore of Prince Regent to avoid the risk of foundering on the many shoals towards Stave. On clear days, there is only the swift current to deal with, but at night or in fog, the tricky channel isn't quite as easy. There was a lighthouse and lightkeeper's cabin on the point at the foot of Prince Regent Island until 1935, when the cabin was towed across the ice to the mainland.

A riverman from Gananoque by the name of Tom Glover was the last lightkeeper at the Narrows. Also in his care were the lights at Jackstraw Shoal and at Lindoe, just off the head of Ash Island out from Ivy Lea. Glover rowed from light to light, filling the early whale oil and then the later kerosene lanterns. During storms, he took shelter in the Prince Regent cabin. He took on a few other chores upon himself as well. He ran an informal library from the cabin, loaning books to islanders and river people. Glover was also the caretaker at the American Canoe Association camp on neighboring Sugar Island, where he would set up the campsites and looked after the wood, water, and other necessities of a camper's life.

Brophy's Lighthouse, on the north shore of Wolfe Island across from the head of Howe Island, is the picture of what one imagines a lighthouse to be, with tidy, bright white buildings and red roofs. The buildings date from 1874 when the government expropriated the property on Knapp's point from the farm of a W. Brophy.

The Rock Island light is on the south side of the American channel, off the head of Wellesley Island. It was built in 1853 to guide ships into the upper section of the ever-narrowing and swiftly flowing channel. The most famous lighthouse keeper in the Thousand Islands was Rock Island's William Johnston. 'Bill', as he was known to all, was a scoundrel, pirate, or folk hero, depending on how you looked at him, who ended his days by serving out a sentence for his role in the Patriot's War with a little lighthouse-keeping duty at the Rock Island Light.

A LOST CHANNEL LEGEND

There are so many small islands and channels in parts of the Thousand Islands that one can easily become lost, especially in the days before landmark buildings and bridges were built. In fact, there is a 'Lost Channel' in the heart of the Thousand Islands, which found its way into folklore in the 1700s. The story of the Lost Channel is based mainly on fact, but requires a little bit of conjecture. During the summer of 1760, a major confrontation was planned whereby the British would cross the east end of Lake Ontario and sweep down the St. Lawrence River with the intent of capturing Montreal and all of the French fortifications along that route. This was the largest force that has ever sailed the river: 10,124 men and their equipment transported in 177 bateaux, 72 whaleboats, and numerous small supply boats, all with the support of two sailing ships, the *Onondaga* and the *Mohawk*. The force left Grenadier Island just southwest of the entrance to the St. Lawrence River on the morning of August 14, 1760, and began making its way down the river. Not long after, a lookout spotted a French bateau put into the stream. Fearing that the French crew would pull ahead and warn the defenders at Fort Levis, just east of present-day Prescott, Captain Loring set out in pursuit in the *Onondaga*. The chase lasted for hours, passing by Wolfe, Grindstone, and Wellesley Islands. Somehow, the bateau always managed to stay ahead, in sight but out of range.

As the day drew to a close and the bateau pulled into the myriad of islands along the north shore of Hill Island, it suddenly became clear to Loring that they had been led into an ambush. Once the *Onondaga* was in the sheltered channels, wind was lost to the sails and the ship's control was lost to the strong pull of the current. There was nothing the crew could

The sun sets on a channel in the Thousand Islands.

do except hope that the ship would not run aground or be hung up in the overhanging trees. When the *Onondaga* was spinning down the narrow passage between Constance and Georgina Islands, the ambush was sprung. Muskets cracked and arrows whistled from the cover of the island forests. Volleys of musket and cannon fire from the *Onondaga* raged back at the French and their Indian allies. Loring managed to get two of the ship's yawl boats into the water. One was to lead the Onondaga free of the channels and the other was to row back upstream to warn the *Mohawk* of the trap that lay ahead. The *Onondaga* somehow managed to avoid being swept into a cliff face and got to the open water and safety beyond the clusters of islands.

The boat and crew who were to have warned the *Mohawk*, however, were never seen again. Loring sent a party back into the maze of islands to find the missing boat, but had to report back that they were unable to find what they thought was the correct passageway, let alone the yawl boat. In his report on the incident, Loring named this place 'The River of the Lost Channel'. There is another story, though, which says that two or three years after the skirmish, the crew of a passing bateau found the sunken remains of a boat with a nameplate that bore the inscription of the ship, the *Onondaga*.

BATEAU CHANNEL

Following the river from west to east, the Bateau Channel lies between Howe Island and the Canadian mainland. Because of the size of Howe Island, this channel is among the longest and most sheltered in the region. The Bateau Channel owes its name to having been the safest route for that workhorse of a craft, the bateau. Bateaux were heavy cargo-carrying rowboats used by French and British military and settlers in making the passage upriver from Montreal to destinations along the river or on the Great Lakes. The river is quite wide and windswept west of the Admiralty Islands, off Gananoque and between the two big islands, Howe and Wolfe. On a blustery day, rowing open and heavily laden boats on this stretch of river could be both wet and dangerous. Even today, the shelter of the Bateau Channel is the choice route for boats under power.

The landscape along the Bateau Channel is quite pastoral in comparison with that found further east in the Thousand Islands. The rugged Precambrian rock of the shield pokes up only here and there through the much more level layered sedimentary bedrock of the mainland and Howe Island shores. In most places, the land tapers gently down to the water. In viewing the farmland, stone houses, and old barns, one can imagine that this placid stretch of river must have been prized by the early Loyalist settlers who drew title to it.

MILLIONAIRE'S ROW

Millionaire's Row is a local name given to the stretch of river that runs along the American shore from the town of Alexandria Bay to the Thousand Islands International Bridge. The name has its roots in the late 1800s, when this region was in its glory as the premier resort area of eastern North America. Some built elaborate, often whimsical and sometimes palatial summer homes from which their well-to-do families could properly enjoy the Islands. This channel is popular on everyone's tour of the Islands today. Many of these summer homes survived the seemingly frequent fires of the early 1900s and are still in regular use.

Castle Rest, George Pullman's summer retreat near Alexandria Bay.
(The Thousand Islands, *James Bayne Company*)

WANDERER'S CHANNEL

Wanderer's Channel winds through the Admiralty Islands, just south of the town of Gananoque. It is one of those wonderful passages among the islands that is deep enough so that even sailboats and tourboats will never touch bottom, yet it is delightfully narrow enough to bring boaters very close to its shores. Wanderer's is a favorite for those who want any excuse to slow down and view the quaint cottages or the swirls of currents and boast of close encounters with shores and shoals.

NEEDLE'S EYE

One of the shortest channels in the region is Needle's Eye, which lies at the northwestern tip of Hill Island, between Hill and two small islands just offshore. The name comes from the days when captains of the steam-powered tourboats while rounding into this narrow passage with its steep shores, overhanging trees, and whirling currents, would boldly announce to their passengers that they dared to thread the Needle's Eye. In so many ways, this little channel epitomizes the character of the Thousand Islands. The strength of the river here, the nearby fragrance of the forests, the pinks and browns of granite bluffs, and picturesque cottages tucked in on shore are the essence of the Thousand Islands, all experienced in a quick dash through the Needle's Eye.

Needles Eye west of Brockville.

The Castanet *and* New Island Wanderer *(opposite) in the Needle's Eye at Hill Island's west tip.* (Les Corbin Collection)

FIDDLER'S ELBOW

Fiddler's Elbow is the name given to a dog-leg channel just north of Bratt, Himes, and Wallace Islands, south of Wood and Ash Islands. While one might think the passage got its name from the bend in the channel, the truth actually makes a better tale.

By the early 1880s, there were a number of steam-powered tourboats on the river, and the competition for passengers among them was becoming inventive. Elisha W. Visger was one of the first tourboat operators in the Thousand Islands in the years after President Grant's historic visit. Visger mapped potential routes for excursions and had a small passenger boat, the *Cygnet*, in service by the mid-1870s. This venture was so successful that by July 879, a larger boat, the *Island Wanderer*, was making regular trips, and he added several more boats to his fleet. Always looking for new ways to promote his tours, Captain Visger, hired a local boatsman, Chauncy Patterson, in the summer of 1881 to play tunes on his fiddle from the shore for the delight of the passengers aboard the *Island Wanderer*. Visger had not chosen just any fiddler, but one who was blind, which added even more to the sensation of the passengers, who considered it remarkable that a blind man could find his way out to the islands and climb up to the top of the cliff face at the west end of Wood Island to play solely for their pleasure. That bend in the channel soon became a formal attraction of the tour and went into Captain Visger's promotional literature as 'Fiddler's Elbow'.

The name of the passage long survived fiddler Chauncy Patterson, though. At about 8:30 at night on August 22,1890, Patterson and his 17 year-old son-in-law set out from Alexandria Bay to cross the river. The steamer *Jaunita*, making its regular run from Cape Vincent to Alexandria Bay, did not see Patterson's unlit skiff in the dark and ran it down. The lad was able to jump from the skiff in time and was later picked up by the *Jaunita's* crew, but the old fiddler was drowned.

As you travel down the Fiddler's Elbow passage, heading east between Ash and Wallace Islands, either in your boat or on a tourboat, the cliff from which the fiddler played looms up like a wall at the end of the channel. It's only when you reach the east end of Wallace Island that you see the turn in the channel. Imagine how precarious a bend this must have been for the freighters that used this channel all those years before the St. Lawrence Seaway was blasted into existence in the late 1950s.

SMUGGLER'S COVE

The channels of Smuggler's Cove figure in plenty of stories about smuggling liquor through the Islands back in the days of Prohibition, but to this day, nobody really wants to talk about them too freely. The stories that do come to light are related in confidence, with names and dates usually left out. Those were hard years here in the Islands, just as they were everywhere in the 1930s, and many an otherwise honest farmer, guide, or townsman could use the quiet cash earned for the odd late-night river crossing.

Word would get to an individual — nobody is saying just how — that a certain cargo needed to be transported to the south shore. It's only a romantic notion that such passages were made Hollywood-style in high-powered, gleaming mahogany launches that sped across the river on moonless nights, slipping into darkened boathouses where the clinking cases of contraband were hastily reloaded into sleek black cars, guarded by gun-toting hoodlums, which rushed the booze to clandestine big city bars. This was not frequently the case. In fact, if there was any common denominator in the stories, it was the ever-so-humble burlap bag.

The burlap bag, the kind used for potatoes dug on the farm, made an ideal container for bottles of spirits. They would stash well in odd-shaped hiding places, and since the bags wouldn't fall apart in water, they would transport well in the bilges of leaky boats. Some of the shipments may not have been made shore-to-shore but rather were set in the shallow water on any of the myriad of rocky shoals that dot the river. The drab brown bags would lie undetected until picked up by another boat, no doubt on a 'fishing expedition'. If the smuggler was discovered and the bags had to be thrown over the side of the boat to get rid of the evidence fast, the burlap could later be snagged on hooks and ropes and be recovered, by one side of the law or the other. The sturdy bags could be slung by rope and pulley across narrow channels, hauled amidst loads of hay or potatoes, or be surrounded by other burlap bags full of duck decoys by a hunter on the river in the fall. But all that is just hearsay, and could be simply the product of an inventive mind. And Smuggler's Cove is just a colorful local name.

INTERNATIONAL RIFT

Except for a very narrow channel called the International Rift, Hill and Wellesley Islands would have been one land mass. The west end of Hill Island nearly joins with Wellesley Island about half way along the latter island's north shore. The St. Lawrence pours vigorously through a rock and cliff-lined passage to empty into the Lake of the Isles. Because the channel runs between the big islands, one owned by the United States and the other by Canada, it is appropriately named the International Rift. Because the relationship of the two countries is so good here in the Thousand Islands, it might be said that this is the only real rift between them.

MOLLY'S GUTS

No one is really sure who Molly is, but she has a couple of 'guts' named after her. A gut is another name for a channel between lands or islands. One of the Molly's Guts runs between the islands of Hickey and Stave, in Canadian waters a little south and east of Landons Bay; the other is about five kilometers west of Brockville, in the area known as Hillcrest, between the mainland and De Watteville Island. Both are shallow but scenic passages, and both are especially important for wildlife. No motorized craft should ever go through the guts because of the disturbance to the habitat of fish, amphibians, and waterfowl. These are like nurseries for the minnows and birds hatched here. Personally powered craft are the proper way to explore these treasures.

LOVER'S LANE

There are plenty of romantic passages amongst the Islands, but the only one that made it to the official charts is Lover's Lane, found between Ash Island and the string of wee islets along its north shore. This narrow and winding waterway, almost hidden from view from the river, is far too small to be negotiated by anything except canoes and rowboats. With the sheltered water overhung by spreading boughs of pines, hemlocks, and maples, Lover's Lane is truly an intimate passage.

BROCKVILLE NARROWS

Compared to other channels in the Thousand Islands, the Brockville Narrows seems anything but narrow. However, from the perspective of the big ships, the 'lakers' and ocean-going vessels that must negotiate this swift water, the passage often seems a little tight, especially so in the summer months when so many pleasure boats traversing these narrows at all rates of speed. The ships seem to almost fill the channel and come so close to the shore that it seems you could almost reach out and touch them. There is no mistaking where the Brockville Narrows goes through the islands. The passageway is so very straight that it seems the islands have been purposefully lined up along both

South end, Lover's Lane.

The International Rift (opposite), as seen in the late 1800s.
(The Thousand Islands, *James Bayne Company*)

Looking west up the Brockville Narrows.

sides. Actually, when the St. Lawrence Seaway was being built, there was some blasting done here to ensure that the freighters would have a cleared route through this island group. This is one of the few ways to get through this section of the Thousand Islands. The waters south of the islands that line the Narrows are strewn with shoals and need detailed local knowledge to traverse safely. To the north of these islands, there are several narrow channels along Brockville's western waterfront. On sunny weekend afternoons, small boats ramble through here as people take in the scenery, pushing up the strong current at Swift Waters and going perhaps just a little further west to thread another Needle's Eye that lies between McDonald Point and a big rock offshore.

THOUSAND ISLANDS INTERNATIONAL BRIDGE

More than just a means of crossing the water to the other side, bridges are often symbols of friendship and commerce. How well this applies to the Thousand Islands International Bridge between Canada and the United States at the heart of the Islands. There is an official but invisible boundary line that weaves through the islands, but so far as Thousand Islanders are concerned, the relationship between the two countries is as if the border was just a formality. The Thousand Islands International Bridge reaches shore-to-shore like a handshake between two friends.

While businessmen in the 19th century often dreamed of building a bridge across the St. Lawrence River in the Thousand Islands, the stonemason's techniques of bridge-building were no match for the power and breadth of the river. Even at the place where the massive islands of Hill and Wellesley nearly form a barrier to the river, there is still a total of almost a kilometer of water to cross. There are a half-dozen channels between the islands and the mainland shores. Through them pours virtually a fifth of the world's fresh water, at practically a million liters per second. The river could never be spanned until engineers had the tools of concrete and cable and riveted steel, and until there was a willingness for the two nations to commit to financing such a project.

Sometime in the 1890s work began on a bridge between Brockville in Canada and

Cars line up on the American shore to cross the Thousand Islands International Bridge for the first time. (Les Corbin Collection)

Morristown in the United States, but it didn't get very far. The plan was to span the river from town-to-town by using the Three Sisters islands as stepping stones where footings would be built. A lack of money, planning, and diplomacy — the bridge would have come ashore in Brockville at the homes of several prominent citizens — prevented any serious progress being made, and the project fizzled. All that remains of the effort is an arch of stone blocks, visible from the shipping channel on the north shore of the Sister Island closest to the Canadian shore.

The idea for the Thousand Islands International Bridge was born in the early 1920s. Ferries made regular river crossings between most of the communities in the area, but these could only operate in the summer months. By the 1930s, the automobile was a fact of everyday life, and although car ferries ran as early and late into the seasons as they could, demand for year round transportation steadily grew. The Thousand Islands became a preferred site for building an international bridge since many prominent politicians had vacationed here and the region held fond memories. While it took several years for the bills to be passed in Canada and the United States that would allow the financing for the bridge project, the way was finally cleared during the Depression years in 1937.

On April 30, 1937, groundbreaking ceremonies took place at Collins Landing, New York. Several thousand people turned out for this occasion marking a turning point in the way of life of the Thousand Islands. Once the bridge was opened, there would no longer be need for the ferry boats and wharves or the people who served them. While the bridge would certainly speed up river crossings and make travel very convenient, it would also mean the end of the socializing and shopping that went on in the various communities that once hosted ferry services.

Firms from both New York City and Montreal were hired by the Thousand Islands Bridge Authority. The engineering project was complex because several types of bridges would be needed to span the river. Foundations for the bridges were started in both countries in May 1937 and were completed by that December. Steel work was begun even as the foundations were being finished. In March and April 1938, cables for both suspension bridges were strung. The steel work for the American span was done in June 1938 and finished for the Canadian spans by that August. Only 16 months after the first work was started, the entire bridge system was complete, 10 weeks ahead of schedule. The project had used 20,700 yards of concrete, 798 tons of reinforcing steel,

At the opening ceremonies of the Thousand Islands International Bridge.
(Les Corbin Collection)

6,500 tons of structural steel, 555 tons of cable, 30,000 barrels of cement, and had taken 575,000 hours of labor. The total cost for the project was $3,050,000 USD.

The portion of the bridge system that actually straddles the border between Canada and the United States is a concrete arch structure, some 27 meters (90 feet) in length, and is faced with local granite. There was originally only one arch in this section of the bridge. This had to be twinned in 1959 to handle the growing amount of traffic.

The Canadian span of the Thousand Islands Bridge is the most elaborate. It involves a 183-meter (600-foot) Warren Truss to bridge the gap from Hill to Constance Island, a 106-meter (348-foot) steel arch from Constance to Georgina Island, and a 229 meter (750-foot) suspension bridge 37 meters (120 feet) above the river from Georgina Island to the mainland of the river's north shore, just east of the Village of Ivy Lea. Overall, the suspension portion is 1,006 meters (3,330 feet) between abutments. The American span has the longest and highest suspension bridge, with a length of 244 meters (800 feet) and a clearance of 46 meters (150 feet) above the river. It spans 1,372 meters (4,500 feet) from abutment to abutment.

The dedication ceremonies for the new bridge were among the biggest events ever held in the Thousand Islands. On August 18, 1938, over 25,000 people were on hand as Prime Minister Mackenzie King and President Franklin D. Roosevelt declared the Thousand Islands International Bridge open. In those early years, about 150,000 cars and trucks crossed the bridge annually, but now those numbers exceed 2,000,000.

In a sense, the work on the bridge is never complete since it requires a great deal of maintenance each year. The bridge must be kept painted and surface work is ongoing. In fact, more than one man has made a lifetime career out of painting the metalwork of the bridge. The demands on the bridge in terms of traffic and loads is well above what would have been imagined in the 1930s, and the bridge must always be kept in

Bridges come in every size in the Thousand Islands.

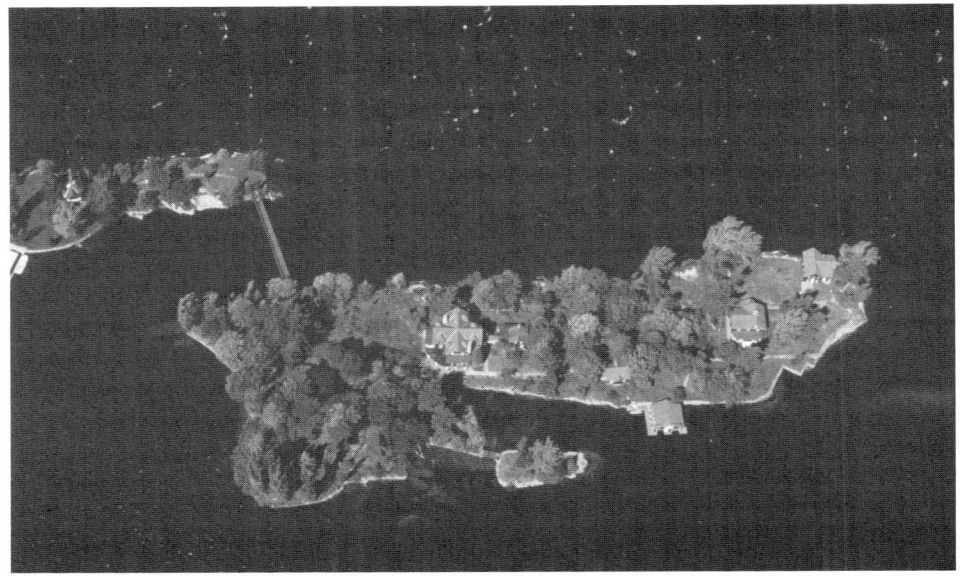

the best condition. The bridge is operated jointly by the Bridge Authority and the St. Lawrence Seaway Authority in Canada. The tolls collected from vehicles allow the bridge system to be operated without having to draw on any public funding.

While the Thousand Islands International Bridge is certainly the most impressive bridge in the region, here and there, small bridges link neighboring islands. A cottage may have been built on one islet and a guest house or boathouse on another.Built on a ore human scale, these often whimsical structures loft from shore to shore to let their owners enjoy a walk to their multiple land holdings. Some of the spans are but a jump. Others are far enough apart and high enough that small boats can navigate under them.

THOUSAND ISLANDS PARKWAY

Closely related to the story of the bridge is that of the Thousand Islands Parkway. When the plans for the bridge were drawn, there actually weren't any major roads that would lead to it. Highway 2, a very old road, runs east-west a few kilometers inland linking communities big and small across southern Ontario. Locally, the highway was crossed by a number of small roadways that led down to villages and their docks. The closest was the route down to Darlingside, a riverside store and steamer fueling depot just east of the site of the bridge. The only other road in the area was the Old River Road, which rambled along parallel to the river from the area of Jones Creek to about Landons Bay. There was really nothing that could service the traffic that would come to and from the bridge.

Brockville's George Fulford, the Liberal member of the provincial legislature, was inspired by the potential boost to the economy that the proposed bridge would bring. Even before bridge work began, he convinced the province to fund a riverfront highway from Gananoque to Brockville. The 'Scenic', as it was called, was a considerable project in its own right. It was also a boon to area farmers and tradespeople who benefited from the work the road provided. But it was far from easy work. Windows rattled time and again throughout the region from heavy blasting to level cliffs and to create rock fill for marshy sections of the route. The once-tranquil woodlands echoed to the thud of axes and the roar of truck engines.

Originally there were four lanes planned for the road, and these were close to being built, which explains why there are the 'extra' spans of bridges at Jones Creek and Landons Bay. However, expenses were running higher than expected because of the terrain, and there was also a fair amount of objection to the project itself because of the upset to the local way of life. Some buildings had to be taken down or moved, and some properties were severed by the roadway. Here and there one can see results of property negotiations that resulted in Parkway features, such as the horse bridge overpass at the Sifton estate near Jones Creek and the underpass at Darlingside.

THE ST. LAWRENCE SEAWAY

Jacques Cartier's second voyage of discovery to North America in 1535 was filled with hope. On his first trip to the New World, he'd found the mouth of the great river that, according to the stories of his native guides, led to faraway lands and perhaps other seas. In this exciting age of exploration, the dream was to find a Northwest Passage across the Atlantic Ocean and through the Americas that would lead to the riches of China and the Far East. Cartier negotiated the lower part of the river, encouraged by the depth and sheer size of the waterway, but noted apprehensively that the water gradually became less and less salty the further inland he came. Then, 1600 kilometers from the sea, he came to the Lachine Rapids — a huge, tumbling flow of such power that no ship could ever pass. From the natives, he learned that there were several more rapids ahead, and it would be a very long journey through freshwater lakes until the source for the river could be found. Now certain that this was not the passage to the East, Cartier turned for home.

The same sets of rapids would challenge all who sought to profit from commerce in the heart of the continent for the next four centuries. While there was no route to spices and silks of the Orient, it became ever clearer as the continent was settled that there

The scale of earthworks for the Seaway construction was staggering.
(Antique Boat Museum)

were great riches in the middle of the continent. Grain from prairie farms had a world-wide market but not an easy route to the sea. The vast iron deposits at the head of the lakes helped build an industrial base along the Great Lakes which manufactured products for markets around the world.

Time and time again plans had been drawn to overcome the wild section of river from Montreal to the east end of the Thousand Islands. Several canals were built in attempts to bypass the sets of rapids. The first were little more than ditches, but did allow safe passage for small craft such as bateaux and Durham boats. But by the time each canal was dug or enlarged, bigger and bigger ships were brought into the Great Lakes trade, making these passages obsolete. Then came competition in the transportation industry from the Erie Barge Canal, a two-meter-deep waterway that provided passage for inland trade from the ice-free port of New York City via the Hudson River and Mohawk River to Buffalo. This canal was shallow and the cargoes of grain and coal had to be loaded and offloaded, but for many years it was a better route than the St. Lawrence. Then came the railways, capable of carrying huge shipments year 'round. Until well into the twentieth century, it seemed that a competitive St. Lawrence canal system would never be built.

But two scenarios would change the course of events — and the course of the St. Lawrence. The lower Great Lakes were well established as the industrial core area of the continent. Iron was brought down from the head of the lakes, and coal, for power and smelting, was shipped up from the Appalachians. Gradually, the iron was running out, but then discoveries were made of huge deposits in Labrador. The only economical way to get the ore inland was up the St. Lawrence, but the canals were far from adequate in size and depth.

Freighter down bound in the Seaway off Alexandria Bay.

While the need to ship Labrador ore alone may not have been enough reason to spend the billions it would take to upgrade the canals, a second incentive came into play. Electricity was generated mainly in coal-fired plants throughout eastern North America. Coal had been a cheap source for power generation, but the technology existed to use the more efficient means of water power. Several big industries saw the potential of the rapids on the St. Lawrence to spin turbines, generating millions of kilowatts. The idea was so attractive that some of them offered to build the dams out of their own finances — an idea that was soon put down by governments on both sides of the river, who didn't want to lose out themselves. Between the need for electricity and the potential to expand industry, the movement to build the St. Lawrence Seaway became an irresistible force.

During the Depression of the 1930s and the war years of the 1940s, there were things other than canals to occupy the minds of governments. Even at the best of times, however, it seemed like the go-ahead for the Seaway would never come. Finally, in the early 1950s, everything pointed to the logic of the enterprise: the need for a deep-water shipping passage, the opportunity to generate enormous amounts of power, and the understanding that if there were tolls placed on the users — the shipping companies — the system might actually pay for itself.

Those were the days when the cost of such big projects was measured almost entirely in dollars alone. It is hard to imagine that a project of this scope would ever be considered today. The impact on the environment, the drowning of farmland, and the uprooting of whole communities would today doom such a project before it got further than the dream stage. Back then, such concerns hardly got any press coverage. To see the greatest impact of the construction of the canal system, look to the section of river between Prescott and Montreal. The new locks, dams, and canals, and especially the 'lost villages' and relocated towns, such as Morrisburg, and the historical site at Upper Canada Village, are the obvious effects of the canal building. In the Thousand Islands, one has to look a little closer. The navigational charts give the best information. In tracing along the line of dashes that marks the path of the shipping channel, one sees dotted lines here and there parallel to the channel where the water depth suddenly changes. Particularly in the Brockville Narrows and to the west of Alexandria Bay, the channel had to be cut through the islands. The dotted lines and abrupt change of depth mark where the interfering shoals and islands were blasted out of the way. In the process, some shoals were heaped higher with broken rock and became large enough to build cottages upon, while a few others ceased to exist.

Before the dams were built at Iroquois to control the outflow of Lake Ontario for power generation and to control water depths for the Seaway, the water levels of the St. Lawrence fluctuated considerably from one season to the next. While there still is a meter of variation from summer highs to winter lows, the level is more constant than in the past. This has allowed more stability in the growth of shoreline aquatic communities and given boaters a fairly predictable range of water levels. Perhaps the greatest change to the life on and in the river has come in a more subtle fashion. The natural 'flushing' action of the lake and river is less than it used to be, and the effects of this are seen as the amount of nutrients has built up in the lakes, dumped there by the growing population around the shores. There has also been the quiet invasion of other plant and animal species into the Great Lakes communities, brought by ocean-going vessels. Many of the weedy species of plants are from the rivers of Europe. The spiny water flea and the ruffie, also from Europe, have upset fish populations. The zebra mussel has become so numerous that as it filter-feeds it stores huge amounts of nutrients and toxins. This has made the water much clearer water, a benefit to recreational divers, but at the same time, this great change in light level and altered nutrient flow has made massive changes to the ecology of the river.

Waiting out the fog at Crossover Light anchorage.

AFTERWORD

International Biosphere Reserve

A memorable view from the heart of the Thousand Islands.

Because of the unique and rich ecology of this region, it is fitting that the Thousand Islands and the southern Canadian section of the Frontenac Arch are being nominated for the status of an International Biosphere Reserve. Yet despite the central military, cultural, and economic role the St. Lawrence River has played in the history of Canada, it has not as yet been designated as one of the Canada's Heritage rivers. The Thousand Islands hosts a phenomenal diversity cf plant and animal life in a landscape that is entirely unique on this planet. The history this most important river corridor to the heart of the continent runs as long, as deep, and as powerfully as the great St. Lawrence River itself. Both designations are truly deserved, but what would such recognition mean?

Pride and honor are the cornerstones of any healthy and successful community. Everyone who lives year round or seasonally or who just spends moments of time in the Thousand Islands has a warm feeling for its special character. When the community of the Thousand Islands collectively realizes the importance of this region and its place in the world, the biggest step will have been taken. To be a Biosphere Reserve and a Heritage River takes nothing away from what the community can do in its continuing development, but it nurtures and celebrates its heritage through wise planning for the future.

Fortunately for us all, there are dozens of park areas in the Thousand Islands. Throughout the Islands and along both shores of the river, there are all manner of parks from little pull-offs and lookouts to those that preserve natural and historic features. In fact, some of the parks in this region are among the oldest on the continent. Even in the days when speculators were going all out to sell real estate to eager summer visitors, and when it seemed likely there would be no public places left to enjoy, movements were afoot to set some of the precious lands aside. Some of the oldest state parks in New York are found along the river as is Canada's first proposed national park.

As in special places throughout the world, the Thousand Islands has several non-government groups and organizations that strive to nurture the character and qualities of the region. Each is motivated to preserve and protect historical, cultural, and ecological features that have attracted their passionate interest, for the collective benefit of all who live or visit here. Their mission statements and contact information is given here should you want to know more about their role in the region. There are many other groups and associations, not to mention individuals and families, that play an important role in preserving and conserving the Thousand Islands, including historical societies, museums, friends associations, cottager associations, and forestry committees. Regrettably, there is insufficient space to include each and every one.

Discovering the Thousand Islands will have fulfilled its purpose if you take even a small part in stewardship of the heritage of the Thousand Islands. If you do, others, decades or even centuries from now, will discover those pine-fragrant breezes that tease the senses across the sparkling river and those tree-framed island heritage scenes that fond memories are made of. History will record your answer.

THE ST. LAWRENCE ISLANDS NATIONAL PARK

St. Lawrence Islands National Park was the first national park proposed for Canada, even though it was not officially reserved until 1905, some 20 years after Banff in the Canadian Rockies. While this is the smallest of Canada's national parks, the holdings are dispersed along the length of the Canadian Thousand Islands, from Cedar Island just east of Kingston to Stovin Island near Brockville. The park has only a small parcel of mainland property, at its headquarters at Mallorytown landing. St. Lawrence Islands National Park plays a vital role in preserving island ecology, while providing places for visitors to explore the living treasures of the Thousand Islands.

St. Lawrence Islands National Park
P.O. Box 469, Mallorytown, Ontario, Canada K0E 1R0

THE ST. LAWRENCE PARKS COMMISSION

The St. Lawrence Parks Commission operates a number of parks and sites, from Adolfustown west of Kingston all the way to the Quebec/Ontario provincial border. Its role is to promote tourism in the region and to protect natural lands and historic features that fall within its jurisdiction. Important to the Thousand Islands are Historic Old Fort Henry at Kingston; Upper Canada Village at Morrisburg — a village of historic homes, farms, and businesses that were relocated to the site during the construction of the St. Lawrence Seaway; day-use and camping parks at Browns Bay and Ivy Lea, respectively; and the Thousand Islands Parkway. The St. Lawrence Parks Commission plays a vital role in protecting the ecological integrity on its substantial holdings of mainland properties in the Thousand Islands.

The St. Lawrence Parks Commission
RR#1, Morrisburg, Ontario, Canada K0C 1X0

NEW YORK STATE DEPARTMENT OF ENVIRONMENTAL CONSERVATION

The New York State Department of Environmental Conservation manages several small parks and day use areas in the Thousand Islands, primarily on the mainland and on the larger islands, Wellesley and Grindstone. The majority are accessible by both car and boat. The parks on Grindstone Island, at Canoe Point, the site of the early American Canoe Association meets, are the exception and can be accessed by boat only. The mission of the DEC is "to conserve, improve and protect its natural resources and environment, and control water, land and air pollution, in order to enhance the health, safety and welfare of the people of the state and their overall economic and social well being."

The New York State Department of Environmental Conservation
Region 6 Headquarters
317 Washington Street, Watertown, New York, U.S.A. 13601

CANADIAN THOUSAND ISLANDS HERITAGE CONSERVANCY

The Canadian Thousand Islands Heritage Conservancy is a volunteer, non-profit, registered charity dedicated to the protection of the natural and cultural resources of the Thousand Islands, for the benefit of present and future generations, by working with landowners and government agencies to find the best methods of heritage resource protection for the region.

The Canadian Thousand Islands Heritage Conservancy
P.O. Box 266, Mallorytown, Ontario, Canada K0E 1R0
www.1000islands-conservancy.on.ca

THOUSAND ISLANDS LAND TRUST

The Thousand Islands Land Trust is a dynamic non-profit land trust located in Clayton, New York. TILT is dedicated to the protection and enhancement of the scenic, recreational, natural, and historic qualities of the St. Lawrence River and its landscape for future generations.

Thousand Islands Land Trust
P.O. Box 238, Clayton, New York, U.S.A. 13624-0238
www.1000islandsschools.org/ESP/TILT

ANTIQUE BOAT MUSEUM

The Antique Boat Museum preserves the boats, engines, and artifacts of Thousand Island vessels and fosters traditional St. Lawrence River boatbuilding skills through workshops and seminars. With over 150 boats and 300 outboard motors and engines in the collection, the ABM contains one of the largest and most impressive collections of inland recreational boats in the world.

Antique Boat Museum
750 Mary Street, Clayton, New York, U.S.A. 13624

SAVE THE RIVER

Save The River is a non-profit, member-based environmental organization whose mission is to preserve and protect the ecological integrity of the Thousand Islands Region of the St. Lawrence River through advocacy, education, and research.

Save The River
P.O. Box 322, Clayton, New York, U.S.A. 13624
www.savetheriver.org

THE ARTHUR CHILD HERITAGE CENTRE AND THE HISTORIC 1000 ISLANDS VILLAGE FOUNDATION

The Historic 1000 Islands Village Foundation and the Arthur Child Heritage Centre are non-profit charitable organizations dedicated to creating exhibits and preserving artifacts that interpret the unique history, life, and times of the Thousand Islands region and its communities.

Arthur Child Heritage Centre
125 Water Street, Gananoque, Ontario, Canada K7G 3E3
www.1000islandsgananoque.com/achcoo.html

LEEDS COUNTY STEWARDSHIP COUNCIL

The Leeds County Stewardship Council serves the land, lakes, and people of Leeds through volunteers and partnerships that help landowners to help themselves.

Leeds County Stewardship Council
P.O. Box 605, Oxford Avenue, Brockville, Ontario, Canada K6V 5Y8

THOUSAND ISLANDS AREA RESIDENTS ASSOCIATION

The Thousand Islands Area Residents Association is dedicated to preserving and improving the present character of the Thousand Islands area with emphasis on the environment. Founded in 1975, TIARA an association over 600 landowners in the Canadian Thousand Islands, both year round and seasonal, who are concerned about the future of this beautiful area.

The Thousand Islands Area Residents Association
Lansdowne, Ontario, Canada K0E 1L0
www.tiara.on.ca

THOUSAND ISLANDS ASSOCIATION

The Thousand Islands Association is a non-profit, charitable organization made up of approximately 1200 sailors, power-boaters, and cottage owners. Founded in 1934, the organization's membership is roughly half Canadian and half American, with a common interest in keeping the St. Lawrence River a safe and beautiful area.

The Thousand Islands Association
P.O. Box 274, Gananoque, Ontario, Canada K7G 2T8
or P.O. Box 81, Alexandria Bay, New York, U.S.A. 13607

TRAVELER'S RESOURCE LIST

To gain more information on the Thousand Islands or to book tours, contact the following organizations. Accommodation information and details of fascinating side trips are available upon request.

Tourism Associations

1000 Islands International Council	800-847-5263
www.visit1000islands.com	
Ontario East Tourism Association	800-567-3278
www.ontarioeast.com	

Chambers of Commerce

Alexandria Bay Chamber of Commerce	800-541-2110
www.thousand islands.com/alexbay	
Brockville Chamber of Commerce	888-251-7676
www.brockville.com	
Cape Vincent Chamber of Commerce	315-654-2481
www.thousandislands.com/capechamber	
Clayton Chamber of Commerce	800-252-9806
www.1000islands-clayton.com	
Gananoque Chamber of Commerce	800-561-1595
www.1000islands.on.ca/gan	

Kingston Chamber of Commerce	888-855-4555
www.kingstoncanada.com	
Prescott & District Chamber of Commerce	613-925-2812

Parks

St. Lawrence Islands National Park	613-923-5261

(Headquarters at Mallorytown Landing, with visitor facilities on some of its 21 islands.)

New York State Parks	315-482-2593

(Over 15 park sites with various combinations of picnic areas, camping sites, docks, and launch ramps.)

Parks of the St. Lawrence	800-437-2233

(Recreational and historic sites along the river, including Fort Henry at Kingston and Upper Canada Village at Morrisburg.)

Museums

Antique Boat Museum	315-686-4104
www.abm.org	
(Clayton, New York)	
Arthur Child Heritage Centre	613-382-2535
www.gananoque.com/heritagecentre	
(Gananoque, Ontario)	

Brockville Museum (Brockville, Ontario)	613-342-4397	Alex Bay Boat Tours (Alexandria Bay, New York)	315-482-tour
Gananoque Museum (Gananoque, Ontario)	613-382-4024	Antique Boat Museum (Clayton, New York)	315-686-4104
Fulford Place National Historic Site (Brockville, Ontario)	613-498-3003	Gananoque Boat Lines (Gananoque, Ontario)	613-382-2146
Marine Museum of the Great Lakes (Kingston, Ontario)	613-542-2261	Heritage 1000 Islands Cruises (Rockport, Ontario)	613-659-3151
1000 Islands Museum of Clayton Clayton, New York	315-686-5794	1000 Islands & Seaway Cruises (Brockville, Ontario)	613-345-7333
Bellevue House National Historic Site (Kingston, Ontario)	613-545-8666	Kingston 1000 Islands Cruises (Kingston, Ontario)	613-549-5544
Fort Wellington National Historic Site (Prescott, Ontario)	613-925-2896	Long Legg Charters (Alexandria Bay, New York)	315-482-3677
Sackets Harbor Historical Society Museum (Sackets Harbor, New York)	315-646-1708	Parkway Boat Line (Ivy Lea, Ontario)	613-659-4622
Sackets Harbor Battlefield Site (Sackets Harbor, New York)	315-646-3634	Rockport Boat Line (Rockport, Ontario)	613-659-3402
		St. Lawrence Cruise Lines (Kingston, Ontario)	613-549-8091

Touring on the Water

		Uncle Sam Boat Tours (Alexandria Bay, New York)	315-482-2611
T.I. Adventures (Kayak Tours) (Clayton, New York)	315-686-2000		
Sea Kayaking in the Thousand Islands (Just east of Gananoque, Ontario)	613-382-4243		

SUGGESTED READING

Anonymous. *The Thousand Islands and the St. Lawrence River: Views Representative of the Wonderful Beauty and Picturesque Scenery to be found in Thousand Island District.* Grand Rapids, Michigan: The James Bayne Company, N.D.

Boyd, Marion Calvin. *The Story of Garden Island.* Kingston, Ontario: N.P., 1973.

Brown, Jack. *Simon Johnston and the Ships of Clayton.* Mallorytown, Ontario: River Heritage Books, 1988.

Corbin, Les and Verda. *The Visger's World: Two Generations of Steamboats, People and Events Associated with the Early Thousand Islands Tours.* Clayton, New York: Les and Verda Corbin, 1987.

Cuthbertson, George A. *Freshwater: A History and Narrative of the Great Lakes.* Toronto, Ontario: The Macmillan Company of Canada, 1931.

Guillet, Edwin C. *The Pioneer Farmer and Backwoodsman.* Volumes One and Two. Toronto, Ontario: The Ontario Publishing Co. Ltd., 1963.

Haight, Canniff. *Country Life in Canada, Fifty Years Ago: Personal Recollections of a Sexagenarian.* Facsimile edition. Belleville, Ontario: Mika Publishing Company, 1971; originally published in 1885 by Hunter, Rose & Co., Toronto, Ontario.

Jacox, Helen P. and Kleinhans, Eugene B. Jr. *Thousand Island Park: One Hundred Years and Then Some.* Thousand Islands Park, New York: Valhalla Printing Co., 1975.

Keats, John and Ringer, Michael. *The Skiff and the River.* Syracuse, New York: Glundal Color, Inc., 1988.

Leavitt, Thad. W.H. *History of Leeds and Grenville.* Facsimile Edition. Belleville, Ontario, 1972; originally published in 1879 by Recorder Press, Brockville, Ontario.

Lucas, Roger A. *Boldt Castle, Heart Island.* Cheektowaga, New York: Research Review Publications, 1992.

Lucas, Roger A. *Bolt's Boats: Alexandria Bay, New York and the Thousand Islands.* Revised Edition. Cheektowaga, New York: Research Review Publications, 1993, 1995.

Mercier, Gilbart B. *Pleasure Yachts of the Thousand Islands.* Circa 1900. Syracuse, New York: Shipyard Press/Shipyard Museum, 1981. Norcom, Stanley. *Grindstone: An Island World Remembered.* New Cumberland, Pennsylvania: Robert Edwards, Publisher, 1993.

Nulton, Laurie Ann. *The Golden Age of the Thousand Islands: Its People and Its Castles.* Binghamton, New York: Vail-Ballou Press, Inc., 1981.

Nulty, Margaret and Mosher, Doris. *Murray Isle: Thousand Islands, Jefferson County, New York*. Gouverneur, New York: MRS Printing, 1972.

Ross, Don. *St. Lawrence Islands National Park*. Vancouver, British Columbia: Douglas and McIntyre, 1983.

Smith, Susan Weston. *The First Summer People: The Thousand Islands 1650-1910*. Erin, Ontario: Boston Mills Press, 1993.

Speltz, Robert. *The Real Runabouts* V. Lake Mills, Iowa.: Graphic Publishing Co., Inc., 1984.

Sykes, Michael and Sykes, Pamela. *Food and Folklore of the Thousand Islands*. Savor the Flavor of the Islands Volume 1. Gananoque, Ontario: Dove Cottage Press, 1995.

Thompson, Shawn. *River Rats: The People of the Thousand Islands*. Burnstown, Ontario: General Store Publishing House Inc., 1989.

Thompson, Shawn. *River's Edge: Reprobates, Rum-runners and the Other Folk of the Thousand Islands*. Burnstown, Ontario: General Store Publishing House Inc., 1991.

Thompson, Shawn. *Soul of the River: Life in the Thousand Islands*. Burnstown, Ontario: General Store Publishing House Inc., 1997.

Wilder, Patrick and Wilder, Michael and St. Lawrence-Eastern Ontario Commission. *Seaway Trail Guide to the War of 1812*. Oswego, New York: Seaway Trail Inc., 1987.

Ice fog in channels above the Canadian Span of the
Thousand Islands International Bridge.